BRITAIN ON THE EDGE OF EUROPE

Britain's role in Europe is a contentious and topical issue. The debate has many dimensions – political, economic and legal – and revolves around the role the country can play on the international stage.

Britain on the Edge of Europe contributes to this debate. The author discusses the country's place in Europe (past, present and future) and challenges current opinion that trade and prosperity are significantly attracted by its location on the periphery of the European Union. The book explores the impact, economically and politically, of the dramatic reorientation of Britain's trade towards Europe in the post-war period and debates its importance for future trade relations.

Using both original research and a fundamental re-evaluation of established opinion, *Britain on the Edge of Europe* provides a contemporary and challenging analysis of the country's future prosperity on the geographical periphery of the European Union.

Michael Chisholm is Professor of Geography at the University of Cambridge.

BRITAIN ON THE EDGE OF EUROPE

Michael Chisholm

London and New York

First published 1995
by Routledge
11 New Fetter Lane, London EC4P 4EE

Simultaneously published in the USA and Canada
by Routledge
29 West 35th Street, New York, NY 10001

Typeset in Garamond by
Solidus (Bristol) Limited
Printed and bound in Great Britain by
Biddles Ltd, Guildford and King's Lynn

British Library Cataloguing in Publication Data
A catalogue record for this book is available from the British Library

Library of Congress Cataloguing in Publication Data
Chisholm, Michael.
Britain on the edge of Europe/Michael Chisholm.
p. cm.
Includes bibliographical references and index.
1. Great Britain–Civilization–European influences. 2. Great Britain–Relations–European Union countries. 3. European Union countries–Relations–Great Britain. 4. Great Britain–Relations––Europe. 5. Europe–Relations–Great Britain. I. Title.
DA110.C563 1994
337.4104–dc20 94–26674

ISBN 0–415–11920–0
0–415–11921–9 (pbk)

to Judith

CONTENTS

FIGURES

TABLES

PREFACE

Three strands of my life came together in this book. For many years, I have been fascinated by locational problems and the related question of overcoming the friction of distance. It has for long been apparent that as the real costs of transport fall, the relevance of relative location either itself declines, or is transferred from the local to the regional or international scale. Over the years, I had concluded that even at the international scale, at least for most manufactures, the spatial differentiation of costs on account of transfer charges is probably small, even negligible. So it seemed that this was a matter that needed to be reviewed systematically, to determine the nature of the reality. The second strand, closely related to the first, is a basic impatience with mechanistic theory which is not subjected to rigorous examination against empirical evidence. In this case, the idea of cumulative causation has a persuasive appeal if the premises are accepted. However, for more than thirty years I have believed that cumulative causation models have been overstated. It has, therefore, been disturbing to see cumulative causation models accepted as self-evidently right and applied in contexts where their utility might be limited. In the specific case of Britain, cumulative causation theories lead to conclusions about our relationship with Europe which seem to me to be questionable. The third and final strand is my own belief that Britain is a part of Europe, and that our future is bound up with that of the mainland. I have been appalled by the foolish things which have been said and done in recent years in the name of British sovereignty and identity, as if we could turn our backs and ignore our neighbours.

It is inevitable that these predispositions colour the work that follows. I believe, nevertheless, that the evidence and arguments marshalled in the following pages will go some way to providing a balanced, non-polemical assessment of the geographical realities facing Britain, leading to the conclusion that our position on the edge of Europe does not present us with permanent and serious economic disadvantages.

I owe a considerable debt of gratitude to many people over many years. It is a particular pleasure to have two colleagues at Cambridge, Andy Cliff and David Keeble, with whom to share mutual interests and both have been of considerable assistance in preparing parts of this book. Daniel Höltgen, a

Ph.D. student and also at Cambridge, has provided me with helpful data and comment on transport matters, especially for freight. Tony Hoare, at Bristol, has been of considerable help in some of my work on British trade and port data, while Ian Gordon at Reading gave me invaluable guidance in working out some issues relating to gravity models. A draft of Chapter 8 was sent to the following, all of whom provided helpful comments and to whom I am indebted for the trouble they took: Mr G. Dunlop, P. & O. European Ferries; Lord Berkeley, Eurotunnel; Ms S.K. Jones, European Passenger Services; Mr A. Howatt, Kent County Council; and Mr W. Shiplee, Rail-freight Distribution.

Ian Agnew, a cartographer in the Department of Geography, drew most of the illustrations; and Sue Slater coped valiantly and patiently with the typing.

Needless to say, I accept full responsibility for any errors there may be, either of omission or commission.

Finally, I am grateful to Tristan Palmer at Routledge for his patience as publisher, for not harrying me as one deadline after another passed by, and to Judith my wife for her forbearance as I abandoned her for my desk.

Michael Chisholm
Cambridge
May 1994

A NOTE ON TERMINOLOGY

The European Community has evolved in many ways since 1958, not least in the name by which it is known. In the early years it went by the name European Economic Community (EEC), then this was abbreviated to European Community (EC) and then, most recently, European Union (EU). It seemed that it would be pedantic to refer to the six, nine or a dozen countries of the EU by using the terms EEC, EC and EU only for the years for which these terms strictly apply. Throughout the text, therefore, EU is to be understood as including the relevant version of the three terms according to the period in question. In just a few places I judged that this would have been confusing and have therefore referred to the EEC or EC as appropriate.

In many places I have used data for some or all of the present twelve members of the EU, irrespective of whether they have been members of the Union for the whole period from 1958. In this context, the following meanings should be noted:

EUR 6 The six original members of the EU – Belgium, France, Germany, Italy, Luxembourg and the Netherlands.
EUR 9 The above six plus Denmark, Ireland and the United Kingdom.
EUR 12 The above nine countries plus Greece, Portugal and Spain.

The acronym EFTA stands for the European Free Trade Association.

In 1990, East and West Germany were united. Throughout the text Germany is used, signifying West Germany up to 1990 and the united Germany thereafter.

As of 1 January 1993, within the EU, non-tariff barriers to trade, and barriers to the mobility of labour and capital, were supposed to have been eliminated, making a Single European Market (SEM).

1

INTRODUCTION

No man is an *Island*, entire of it self
John Donne, *Meditation XVII*

The deep ambivalence of British attitudes to Europe has been painfully obvious in recent years. Mr Major as Prime Minister has said that he desires Britain to be at the heart of the European Union (EU), and yet in the 1991 negotiations for the Maastricht Treaty he fought for, and obtained, a series of opt-out arrangements which clearly indicate a less enthusiastic view of the Community than is held by other member states. This lack of enthusiasm was reinforced first by the September 1992 débâcle over the Exchange Rate Mechanism (ERM) when the pound sterling ignominiously quit the ERM and devalued, and then by the incomprehensible manoeuvring on the ratification of the Maastricht Treaty which took an inordinate amount of Parliamentary time and energy in late 1992 and 1993. Then again, having pressed for the enlargement of the EU, Britain was not willing to accept the logic of changing the voting rules and tried vainly to insist on maintaining the blocking minority of votes at twenty-three, risking jeopardy to the admission of Austria, Finland, Norway and Sweden, only to be forced into a humiliating climb-down in March 1994. Thus could it be said of Mr Major that: 'We still don't know where his heart really lies' (*Economist*, 26 March 1994: 37). Indeed, it seems quite clear that the European question will continue to trouble the Conservative Party in particular, and the British body politic more generally, for the foreseeable future.

Discussions about Britain's present and future position in the EU need to take account of three sets of recent changes which have major significance. The collapse of Communist control in the former Soviet Union and Eastern Europe has dramatically changed the geopolitical balance of the entire world, and opened up new relationships between east and west Europe. The EU has itself been changing in both geographical extent and in its internal organization. And, third, whereas up to the 1980s the EU was perceived to be a dynamo for economic growth, benefiting both the member countries and the rest of the world, in the last few years that perception has changed sharply;

severe and prolonged recession has been accentuated by high interest rates, occasioned in part by the high cost of uniting East and West Germany. To set the scene for the remainder of the book, these three issues will be briefly discussed.

November 1989 saw the breaking of the Berlin Wall, the beginning of the end of Communist control in the Soviet Union and in Eastern Europe. The great initial hopes have been tempered by the realization that we all now live with an unstable and unpredictable situation in these former Communist countries, such that the very concept of Europe is now back on the drawing board (Crouch and Marquand 1992). Although the Czechs and Slovaks have arranged a peaceful divorce, serious conflict mars relationships between several of the states which have emerged in the former Soviet Union, and Yugoslavia has been torn asunder. Meantime, although some of these former Communist countries seem to be making the transition to market economies, most notably perhaps the Czech Republic, Hungary and Poland, the pain is great and success may be jeopardized if Russia, the Ukraine and other members of the new Commonwealth of Independent States fail to make peaceful changes to their political and economic systems. None of us can be sure of the outcome, beyond noting that whereas the danger of global superpower conflict appears to have receded, savage conflict in Europe's backyard is a present reality which may last for far too long and which may not remain confined to conveniently remote and unfamiliar places. If we travel with hope, we must nevertheless tread with caution.

The EU was formed in 1958 as the European Economic Community, comprising six countries – Belgium, France, Germany, Italy, Luxembourg and the Netherlands (EUR6; see Figure 1.1). Fifteen years later in 1973, Britain, Denmark and Ireland joined, Britain having attempted to join on two previous occasions, only to be rebuffed by de Gaulle. Eight years after the first enlargement, Greece became the tenth member, followed by Portugal and Spain in 1986. After this rapid expansion in the 1980s, the mood was generally in favour of consolidation in the expectation that any further addition of members would take place in a relatively leisurely manner. That expectation has been confounded by events. The helter-skelter union of East and West Germany in October 1990 presented the EU with an unexpected *de facto* enlargement. Just before this happened, talks had been initiated between the EU and the European Free Trade Association (EFTA, formed in 1960); these culminated in an agreement in October 1991, which came into effect on 1 January 1994, creating the world's largest free trade area, known as the European Economic Area (EEA). Maintaining the tempo of change, negotiations began in 1993 for the entry into the EU of Austria, Finland, Norway and Sweden. Subject to the outcome of national referenda, these countries will become members in 1995. The ink was barely dry on these arrangements when Hungary applied to join the EU, joined by other East European countries, so that perhaps by the year 2000 the EU will extend to

Figure 1.1 Member countries of the EU and of EFTA

the borders of the former Soviet Union and marginally beyond in the case of the three Baltic countries of Estonia, Latvia and Lithuania.

At the time of the first enlargement in 1973, the three new members were substantially poorer, in terms of GDP per person, than the original six members of the EU (Table 1.1). With the subsequent addition of Greece, Portugal and Spain, the pattern seemed to be clearly set – that new and 'peripheral' members of the EC are relatively poor. That pattern is about to be reversed. The three Scandinavian countries and Austria all have a GDP per caput which is at or above the EUR12 average; indeed, Norway is richer than any of the EU states. These four countries have achieved this high level of income despite being outside the Union since its inception and despite being every bit as much 'peripheral' as Britain. In fact, it has been a matter of explicit policy that the EU will entertain applications for membership only from countries that are likely to be net contributers to EU funds. This reflects the lack of dynamism in recent years in the EU, in marked contrast to earlier years when the EU was willing to contemplate the collective burden implied by admitting relatively poor countries. If the Union is to be expanded eastwards, there will again have to be a willingness on the part of the richer countries to make sacrifices.

In parallel with the momentous geopolitical changes in the former Soviet sphere of influence and the progressive enlargement of the EU, the EU itself

Table 1.1 Population and GNP per caput, European countries

	Population, millions	*GNP per caput, US$ current prices*	
	1991	*1973*	*1991*
Belgium	10.0	4,990	19,300
France	57.0	4,810	20,600
Germany	80.1	5,690	23,650
Italy	57.8	2,520	18,580
Luxembourg	0.4	5,460	31,080
Netherlands	15.1	4,670	18,560
Denmark	5.2	5,870	23,660
Ireland	3.5	2,150	10,780
United Kingdom	57.6	3,270	16,750
Greece	10.3	1,980	6,230
Portugal	9.9	1,440	5,620
Spain	39.0	2,170	12,460
Austria	7.8	3,900	20,380
Finland	5.0	4,120	24,400
Norway	4.3	5,190	24,160
Sweden	8.6	6,360	25,490

Source: *World Bank Atlas*, World Bank 1993

has been changing. When it was initially formed, the EU took three important steps with the aim of facilitating trade between the original six members. Probably the most important of these steps was the progressive elimination of tariff barriers between the member states, from their initial level of about 12 per cent. The second step was to abolish the practice in setting rail freight rates of treating each national frontier as if it marked the beginning of an entirely new consignment, notwithstanding that freight was in transit across the frontier; the effect was to add artificial increments to the freight charges, being the notional 'terminal' charges incurred in loading and unloading goods at the frontier. Third, the Common Agricultural Policy (CAP) was established, embodying, *inter alia*, the idea of a single price within the EU for the main agricultural products. By general agreement, the CAP has outlived such usefulness as it may ever have had, and some tentative steps have been taken to modify the rules.

In the non-agricultural sector, important steps have been taken to build on the liberalization of trade initiated by the abolition of tariffs and artificial rail freight charges. Agreement was reached in 1986 to introduce the Single European Market (SEM) by the end of 1992. This is an ambitious attempt to eliminate non-tariff barriers to trade and also to facilitate the mobility of the factors of production. Non-tariff barriers to trade include frontier documentation processes, national technical standards and government procurement policies. In theory, all of these trade barriers should have been eliminated by the end of 1992. As for the factors of production, capital movements have in fact been progressively freed from most controls and the main new feature of the SEM is the abolition of border controls on the movement of people, although Britain negotiated an opt-out clause on this issue.

Once it is fully implemented, the SEM will mark a giant stride towards the liberalization of the EU economy. This liberalization has been presented as providing: 'The economic context for the regeneration of European industry in both goods and services; and it will give a permanent boost to the prosperity of the people of Europe and indeed of the world as a whole' (Lord Cockfield 1988: xiii). Summarizing the EU studies of the 'costs of non-Europe', Cecchini (1988: 103) put this benefit at between 4 and 7 per cent of Community GDP. While this claim may be exaggerated, it does illustrate the hopes which are pinned on the consolidation of the Union. But these hopes have their mirror in the fears that are engendered by the progress to closer economic and political union, clearly exemplified by attitudes in Britain to the Maastricht Treaty.

MAASTRICHT

The Treaty of Maastricht re-writes the original Treaties of Rome and marks a further major step beyond the SEM. Although Britain played a constructive role in much of the negotiation, her stance on certain issues left a clear impression that she 'is no mainstream European yet' (*Economist*, 14 December 1991: 29). During a debate in the House of Commons in November, prior to the Maastricht summit, Mr Major had committed himself to avoiding a 'social chapter' on the rights of workers, to limiting any increase in the powers of the European Parliament and Commission, to ensuring that if or when a single European currency comes into existence Britain would not be bound to join, and, above all else, that there should be no mention of the aim to create a federal structure for the EU. He came back from Maastricht able to claim that he had achieved all these aims.

The Treaty establishes the European Union as the successor of the European Community, with the aims of: economic and monetary union and ultimately a single currency; working towards a common foreign and defence policy; European citizenship; and closer co-operation on justice and home affairs. A timetable was set for establishing a single currency no later than 1999, including the establishment of an independent European central bank. With the partial collapse of the ERM in 1992, it is not clear that this timetable is feasible, or even that the aim is achievable; this is somewhat ironic, given that Britain explicitly negotiated the right to opt out of these arrangements.

EURO-SCEPTICISM

The Parliamentary debate which preceded the Maastricht negotiations was: 'A thoroughly insular affair, a debate more about Britain – and the Conservative Party – than about Europe and its future' (Peter Jenkins, *The Independent*, 21 November 1991). The anxieties which were then expressed, and which continue to be articulated by the Euro-sceptics, arise from two sources: first, the concept of sovereignty; and second, the fear that closer union with the EU will result in Britain losing out economically. In practice, of course, these two issues are interrelated. Mrs Thatcher has been adamant that nothing should be done which further limits the power of Parliament and of government, using the rhetoric of sovereignty as her clarion call to resist moves towards federalism and centralized economic, political and social management. This call is based on the belief that the organs of national government can and should have control over the internal affairs of a nation, that external 'interference' is unacceptable; and that nations should be able to exercise sovereign power in foreign affairs. Such a concept of sovereignty harks back to the days when nations were largely self-sufficient in economic terms, even though willing to engage in military and political alliances. If foreign economic transactions are of small moment for a nation, then indeed the government of the day can impose the terms on which individuals and firms operate. However, just as 'No man is an *Island*, entire of it self' (John Donne), so also no advanced nation is an independent actor on the world stage. In economic terms, the more a country's economy is meshed with the economies of other countries, the fewer are the macro-economic and micro-economic powers which are available to governments. If interest rates get seriously out of line with international rates, balance of payments problems will ensue as money moves in or out. If a country seeks to peg its exchange rate above or below the level indicated by the currency markets, there will be repercussions for the balance of payments, interest rates and the level of activity, and hence of employment and unemployment. If business taxes are regarded as too onerous, firms will move offshore. The catalogue would be a long one, and in every case we would have to admit that sovereignty is a very heavily qualified concept, and furthermore that the direction of change is to *reduce* the ability of governments to control economic affairs within their borders in a direct manner.

The issue, therefore, is not between the maintenance of sovereignty and its abandonment; to present it in these terms is unhelpful. The real issue is to work out how the sovereignty of nations should be modified to accommodate the self-evident fact that the nations of the world are becoming increasingly interdependent in all kinds of ways. Finding means for nations to work together is a pragmatic matter, not the abandonment of the principle of sovereignty.

The second fear is that Britain, being a country with an economy that in recent years has performed rather badly relative to others, will lose out if there are further moves to economic integration and the further reduction of control over the economy on the part of government. At the general level, the proposition runs along the following lines. Britain's economy is rather weak, and the rate of growth rather low, because too many firms are insufficiently competitive in the domestic and international markets. Under these circumstances, further integration with the EU (whether EUR12 as at the time of writing, or as it may be after enlargement) will expose British firms to competitive pressures they will find hard to withstand. If, at the same time, the power of government to pursue macro-economic and/or micro-economic policies is diminished, it will be impossible to provide British firms with the domestic circumstances they need to prosper. Therefore, so the argument runs, importers will have a field day and British firms will be even harder pressed to hold their own in domestic and overseas markets. On this analysis, progress towards closer economic union should wait on measures to enhance the performance of the British economy.

That Britain has been plagued by a long history of poor competitiveness is not in doubt. What is much less clear is the nature of the intervention that is required to put matters right. Given the openness of the economy, macro-economic regulation is already severely constrained, and experience from the heyday of Keynesian management is in any case not particularly encouraging. Governments do, however, still have considerable control over important supply-side aspects of the economy, which, if suitably manipulated, should have a beneficial impact on overall competitiveness. Nevertheless, and despite the rhetoric, successive Conservative administrations have not succeeded in creating the supply-side revolution which they have proclaimed since 1979.

For any country, the fact is both simple and brutal: if its firms can match the competitiveness of foreign companies in markets both domestic and worldwide, it will be able to prosper. If significant numbers of firms fail in that challenge, the country will become relatively, and maybe absolutely, poorer. This will be manifest either in high levels of unemployment and the visible loss of potential national product, or in reasonably full employment in a sheltered economy at the price of low output per worker and hence low incomes. To provide shelter against the chill winds of competition could be justified if this were the only means, or the best means, available to achieve the transition from uncompetitiveness to competitiveness. I personally do not believe that this is the general case, though I am willing to accept that specific industries might warrant short-term assistance in closely defined circumstances.

Thus, in 1994, even before the government's inept handling of the question of the number of votes needed to block a proposition in the EU, Britain stood in an ambivalent, 'semi-detached' position – a full member of the Union but the most reluctant in building for the future, the most strenuous in trying to hold on to an outdated concept of sovereignty (George 1990). Until, or unless, these basic attitudes change, Britain will indeed be on the edge of Europe. But it is the nature of attitudes that they can and do change, so there is no immutable reason why hostility to the European enterprise need continue indefinitely – unless, that is, there are permanent and unchangeable factors which put Britain apart from the rest of Europe.

BRITAIN ON THE PERIPHERY?

There is an influential strand of thinking which holds that a country, such as Britain, on the periphery of the EU suffers a substantial economic penalty. The argument goes as follows. When firms make their investment choices, they select a location which will maximize their profits. If other things are equal, they will wish to minimize the cost of access to their inputs – components, design contractors, professional services and so on – and at the same time minimize the cost of reaching customers. To minimize total transfer costs, a location which is 'central' in the economic space is to be preferred. A popular way of assessing 'centrality' is to use measures of economic potential (Clark 1966; Clark *et al.* 1969; Keeble *et al.* 1982, 1988). Economic potential shows the relative accessibility of each area to all other areas within the economic space. Calculated for the EU, economic potential is rather low in much of Britain, with only the south-eastern part of the country reaching a level comparable to the most accessible regions of mainland Europe. Britain as a whole is somewhat peripheral to the rest of the

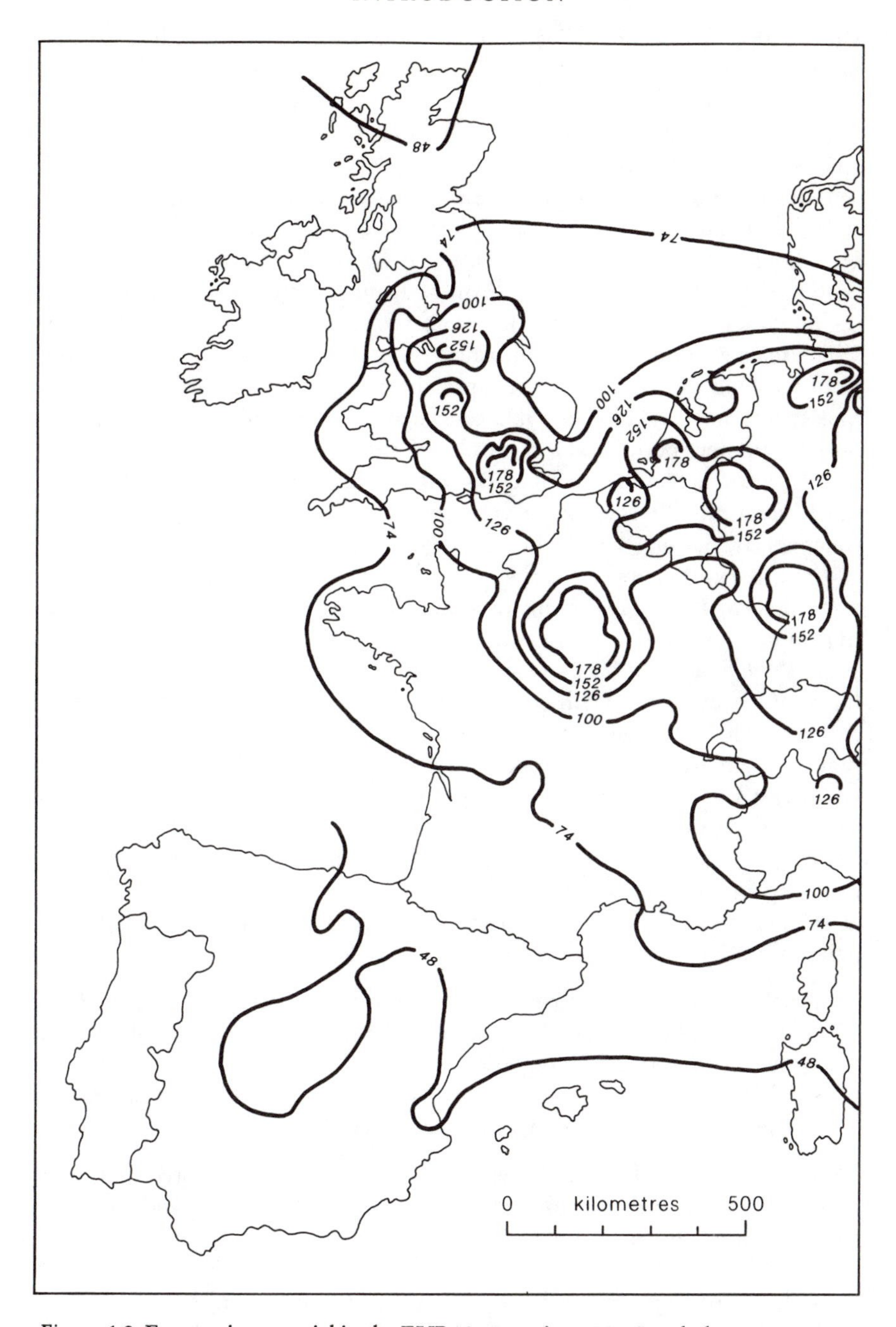

Figure 1.2 Economic potential in the EUR12 countries, 1983. Isopleths as percentage of the EUR12 average, at intervals of half a standard deviation
Source: Keeble *et al.* 1986: 26

Union, and this peripherality is accentuated towards the South West, the West and the North, being most evident in Cornwall, Wales, Northern Ireland and Scotland (Figure 1.2).

The static problems posed by peripheral location are, so it is argued, compounded by two sets of dynamic processes. Of these, cumulative causation is the more familiar. Cumulative growth theorizing denies the spatially equilibrating process identified by neo-classical economists. Hirschman (1958) and Myrdal (1957) start by emphasizing the role of scale economies (both internal to the firm and, more important, externally). A region which gets a head start, for whatever reason, provides conditions in which scale economies can be obtained, which enhances the profits accruing to firms. This will attract further capital investment. At the same time, these more profitable firms located in a central area can pay higher wages than is possible in more peripheral locations, with the result that labour will migrate to the central area. The simultaneous migration of labour and capital in the same direction fosters cumulative growth in the more favoured areas and (relative) stagnation, even decline, in the more peripheral locations.

The theory of economic union draws attention to the trade-creating and trade-diverting effects of reducing import duties, and, *a fortiori*, any other impediments to trade, between a group of co-operating countries (Balassa 1962, 1989; Scitovsky 1958). Geographically, the main benefits to be derived from these trade effects will be along the internal frontiers, and especially in any area where a large number of frontiers is located. Within the EU, the region which most benefits from this effect is centred on Belgium. Although south-east England is immediately adjacent to this favoured area, the rest of the country is rather remote.

The proponents of this general line of reasoning argue that, whatever the overall gain for the EU, Britain stands to lose from the consolidation of economic integration – and also the opening of the Channel Tunnel – in either or both of two ways.

1. At the national level, Britain is perceived to be peripheral to the EU as a whole, and therefore likely to suffer the negative effects of cumulative growth processes and integration effects.
2. Within Britain, it is the south-east which will gain but at the expense of the more remote western and northern regions.

There is a substantial body of literature which takes it for granted that these centralization processes dominate the processes reshaping the geography of the EU. For example:

> With regard to peripherality, there can be little question that much of the 'North' of the UK suffers from its remoteness from London and the South East, so that a shift of the economic centre of gravity further towards the core of the Community would exacerbate this phenom-

> enon. Indeed, it is arguable that the UK as a whole is peripheral in Community terms, even if not so obviously as countries like Greece or Ireland. The main disadvantages of peripherality stem from the higher costs of communications and of market access.... Unless there are sufficiently lower factor costs in peripheral areas, it would be expected that such economic activities would tend to gravitate towards the core of the EC.
>
> (Begg 1990: 90–1)

Despite the caveat, which is reinforced in a later paper (Begg and Mayes 1993), the tenor of this paper is that centralization is to be expected, a view put rather more forcibly in another recent publication:

> The move to a Single European Market is expected to reallocate markets and redistribute production in favour of the most efficient and best situated firms. In addition to greater concentration of industry, it implies growing divergence at regional level. The gains from trade are expected to concentrate at the centre of the Community in the more prosperous regions: transient and long-term unemployment emerge at the periphery.
>
> (Mackay 1992: 278)

Views such as these, concerning the impact of further integration, have become embedded in the conventional wisdom of the EU itself:

> Historical experience suggests, however, that in the absence of countervailing policies, the overall impact on peripheral regions could be negative. Transport costs and economies of scale would tend to favour a shift in economic activity away from less developed regions, especially if they were at the periphery of the Community, to the highly developed areas at its centre.
>
> (Committee for the Study of Economic and Monetary Union 1989: 22)

According to this line of reasoning, anything which reduces trade barriers and impediments to the mobility of the factors of production in the EU will serve to accentuate centralization pressures, to the detriment of Britain as a whole but especially of her peripheral regions. That the fears exist and are real cannot be denied. However, the question to consider is whether there is a sound basis for the fears that Britain's geographical location gives rise to a permanent disadvantage? If there is such a disadvantage, then this may be a part explanation for the weak performance of the British economy. Alternatively, geographical location may have little or no impact on the relative success of the British economy. It is this question – does location on the edge of Europe pose economic problems for Britain? – that is the focus of the present study.

If we may anticipate our enquiries, the evidence which is reviewed in the following chapters points very strongly to the conclusion that Britain's peripheral location is a matter of small importance. Freed from the encumbrances of a mistaken analysis, the way is thereby cleared for a forward-looking approach to two important issues. The first of these concerns the competitiveness of British manufacturing and service industries. As we cannot blame our location for our weak economic performance, we must look elsewhere for the source of trouble and for the remedies. The second concerns the nature of the new and improved transport links that Britain needs to forge with mainland Europe, and the geographical implications of new facilities such as the Channel Tunnel. In the compass of this study it will not be possible to treat either of these topics as fully as they deserve, but at least the ground will be cleared for further investigation.

STRUCTURE OF THE BOOK

The central purpose of the book is to explore the reality that either underlies, or belies, the fears which have been documented that Britain suffers serious economic disadvantages from its position on the geographical periphery of Europe. To begin this exploration, it is essential to consider Britain on the one hand, and Europe on the other, in the international trading system, since our relationship with Europe is but part – albeit an important part – of our total international connections. Chapter 2, therefore, is primarily an empirical review of post-war changes in international trade, focusing on Britain and Europe. For this purpose, Europe is taken to be the twelve present members of the European Union (EUR12) as it existed in 1994. Chapter 3 then introduces one of the bodies of theory that may be useful in explaining the geography of international trade – namely, the gravity model. This model predicts that the volume of trade should be inversely related to the distance (transfer cost) which separates trade partners. However, the limited empirical work that has been done with the gravity model in international trade indicates that distance (transfer cost) is not a very important factor – especially for manufactured goods which now dominate international trade. This finding implies that the geographical location of Britain is probably not a major factor explaining the performance of the economy. That implication runs counter to two other bodies of spatial theory. The first of these, the spatial consequences of economic integration, holds that the formation of an economic union such as the EU will favour countries and regions that are 'central' in the sense that numerous and important barriers to trade have been removed. Chapter 4 sets out the theoretical arguments and then reviews the empirical evidence that is available for Europe for the magnitude of the effects of economic integration. This evidence suggests that the centralization thesis derived from the theory of economic union receives scant support. The second body of centralization theory, known as cumulative causation, is reviewed in Chapter 5 and confronted with the available empirical evidence concerning regional rates of growth. Although there is some truth in the cumulative causation model, its utility is found to be limited.

By this stage in the review, the evidence points strongly to the proposition that relative location may not be a particularly important factor for Britain in relation to Europe. One reason why this might be so is derived from neoclassical economic thinking, which draws attention to the relative mobilities of the factors of production, to differences in the cost functions facing firms, and hence to the role of prices in the allocation of resources between sectors and places. Wage differences in particular indicate that some of the 'peripheral' countries, especially Spain and Portugal, stand to gain considerably from the further integration of the EU that occurred with the putative completion of the SEM on 31 December 1992. This evidence, reviewed in Chapter 5, is followed in Chapter 6 by an examination of spatial variations in transport costs in Britain relative to other costs. Although the north and west of the country may suffer from higher transport costs, the magnitude of this disadvantage is small (at the standard region level maybe even non-existent) and is in any case offset by compensating lower costs for labour and other inputs. Another aspect of the matter is examined in Chapter 7 – namely, the changing geography of commodity traffic through Britain's ports and the passenger traffic by sea and air. As Europe becomes economically more important for Britain relative to the rest of the world, one would expect a reorientation of traffic towards points of entry located in the south-east. In fact, the picture is rather mixed: in the freight sector, there is some shift in the expected direction although simplistic cause-and-effect propositions are not in order; at least as important is the fact that passenger traffic does not show the same reorientation towards Europe as does freight and that the tendency is for more northerly airports to gain market share at the expense of those in the south.

Chapter 8 turns to an examination of Britain's infrastructure needs in the light of the preceding analysis – links with Europe on the one hand, and internally on the other. The potential impact of the Channel Tunnel figures large in this assessment. The Tunnel will probably have a smaller economic and regional impact than its proponents, and others, would have us believe, and the really critical issue is the failure of government to pursue a properly articulated transport investment strategy to link the Tunnel to the national rail system. Finally, Chapter 9 draws the overall conclusion that being geographically 'peripheral' in Europe is a matter of small moment economically for Britain. If our economy is weak, we cannot lay the blame on our relative location but must look inwards to other factors which affect our efficiency and which, unlike relative location, are under our own control.

2

BRITAIN, THE EUROPEAN UNION AND WORLD TRADE

> Britain tends to see itself (and is certainly seen by its partners) as having become a member of the Community in, so to speak, a fit of despair rather than as the result of conviction and enthusiasm.
>
> (Nevin 1990: 20)

> Economists divide notoriously because they look at different facts, or interpret the same facts differently.
>
> (Bhagwati 1991: 99)

To understand Britain's changing economic relationships with Europe, it is essential to consider the rapid evolution of the global economy which has occurred since the Second World War. The broad outlines of this evolution have been succinctly described in the following terms:

> Markets for both goods and financial assets became further globalized in the 1980s as new technologies reduced the economic distance between countries in quantum jumps. The integration of the financial market was particularly dramatic.
>
> After a period of trade pessimism in the early 1980s, world trade grew vigorously in the latter years of the decade. The value of world exports increased from under two trillion dollars in 1980 to over three trillion in 1989. The volume of trade increased by 4 per cent a year. Most of the growth came from trade in manufactures which grew much faster than world output of manufactures. Increased product differentiation, greater intra-industry trade and great advances in communications and transport technologies linked national markets more tightly.
>
> (United Nations 1990: 3)

Certain aspects of these global changes are of particular relevance in the present context and provide the subject matter for this chapter, and also the context for the remainder of this book.

A word must be said specifically about services as distinct from commodities, and why it is that the discussion will focus on commodities. International transactions in services of all kinds have been expanding in volume and value more rapidly than has commodity trade (Corbridge and Agnew 1991; Marshall *et al.* 1988). These services include the financial sector, advisory and consulting services. The rapid expansion of international trade in services has been made possible by the growth of air transport and the revolution recently wrought in telecommunications. The costs of personal communication around the world have been so sharply reduced that the ability of a country to compete in the provision of international services has very little to do with its location and very much to do with the quality of its internal infrastructure and of the services it can offer to the world. Indeed, the proposition has even been made that global financial integration is now so complete that we are witnessing the end of geography (O'Brien 1992). While this claim is exaggerated, the kernel of truth – that relative location on the global scale is not particularly important for the provision of international services – implies that in this sector the location of Britain on the edge of Europe is not a matter of great moment.

As for commodities, though, transfer costs between nations remain significant, despite considerable changes which have reduced the real costs of shipments. A widely used rule of thumb is that, on average, international trade adds 10 per cent to the cost of the goods being traded. There is, therefore, a prima-facie case that the geographical location of a country within the international trading system will have a material impact on its prosperity, especially if, as is the case for Britain, international trade is an important feature of the economy.

It follows that if the geographical location of Britain is a material factor affecting the nation's prosperity, the evidence will be most apparent in the commodity sector. If there is clear evidence that in commodity trade geographical location does have a measurable impact, then it might be worth exploring the service industries to see whether the same is true of this sector. The first task, though, is to examine the commodity sector, in our exploration of the propositions discussed in Chapter 1.

LONG-TERM CHANGES IN THE IMPORTANCE OF COMMODITY TRADE

The long-term trend appears to be quite clear: that the proportion of commodity production which enters international trade is rising, which is another way of saying that economies are becoming more open. Although the trend seems to be well established, there has been – and no doubt will continue to be – considerable fluctuation about the trend.

In 1800, the value of commodity imports plus exports was about 3 per cent of world output. By 1913, this proportion had risen elevenfold to 33 per cent. Most of this dramatic increase in the openness of national economies had occurred by 1870, after which date there was little change until the First

World War. International trade was then seriously disrupted by the two world wars, by the recession of the 1930s and by the measures taken by governments to protect their national economies (Molle 1990). Immediately after the Second World War, protection was still at high levels, with the consequence that international trade as a proportion of world output was initially well below the 1913 level. Throughout most of the post-war period, trade has expanded more rapidly than output, so that by 1988 the sum of imports plus exports was equivalent to 31.5 per cent of output, almost back to the situation immediately prior to the First World War (Chisholm 1990a: 95; World Bank 1990: 183, 205). After 1988, the trade proportion declined quite sharply as recession bit into the economies of the developed countries but nevertheless had recovered to 31.6 per cent in 1991 (World Bank 1993: 243, 265). Most commentators expect the relative importance of trade to continue its long-term upward trend, an expectation which looks reasonable given the successful conclusion of the Uruguay round of GATT in December 1993. Over the whole of the post-war period, successive rounds of negotiation under GATT have served to reduce tariff barriers to trade and, more recently, have begun to tackle non-tariff impediments as well. The GATT framework has been immensely important for liberalizing the international trade regime, and so far, despite the pressures arising from national sectional interests, the political will to continue the process seems to be intact. Therefore, there is a high probability that the long-term trend to greater openness of national economies will continue. However, whether that expectation is fulfilled in the future is of lesser importance in the present context than the fact that, throughout practically the entire post-war period to date, international trade as a proportion of output has been rising. Furthermore, this trend to increased openness has characterized the economies of most countries, including Britain (Grassman 1980; Beenstock and Warburton 1983).

The general trend to more open national economies can be attributed to several factors other than the regime of international agreements which regulates the terms on which transactions occur. Transport improvements by land and sea over the last two centuries have reduced the real costs of moving goods. At the same time, the greater ease of transmitting information – by telegraph, radio and satellite telecommunications – implies greater efficiency in the organization of shipments and hence further cost savings (Chisholm 1979, 1990a; Dicken 1986; Kindleberger 1956; Latham 1978). In addition, technological change has resulted in ever-increasing efficiency in the use of fuels and raw materials to achieve a given end product, and the unit value of products relative to their weight and/or volume has been rising. The net effect of all of these trends has been to reduce the real cost of transport and therefore to increase the scope for gains from trade.

Although at the aggregate level the relative importance of commodity trade has returned to near its 1913 level, this fact masks a structural change

Table 2.1 Volume changes in world exports, averages for periods 1950–65 to 1980–6 (percentage per annum)

	1950–65	*1965–70*	*1970–5*	*1975–80*	*1980–6*
Agricultural products and foodstuffs					
Change p.a. (%)	4.6	4.0	1.2	5.9	1.1
Share of world exports (%)	25.1	19.0	15.4	14.7	15.4
Apparent elasticity in relation to GDP[1]	0.9	0.7	0.3	1.5	0.4
Mineral and energy products					
Change p.a. (%)	7.1	9.3	0.6	2.8	–1.9
Share of world exports (%)	39.8	38.5	35.7	31.0	23.8
Apparent elasticity in relation to GDP[1]	1.5	1.7	0.2	0.7	–0.7
Manufactures					
Change p.a. (%)	9.0	11.5	7.4	6.8	4.5
Share of world exports (%)	35.1	42.5	48.9	54.3	60.8
Apparent elasticity in relation to GDP[1]	1.9	2.1	1.9	1.7	1.7

Note: [1] Ratio of variation in export volume to variation in world GDP
Source: Commission of the European Communities 1989: 14

of great significance that has occurred almost entirely since the Second World War. During the nineteenth century and the first half of the twentieth, primary commodities accounted for three-fifths of total trade; the remaining two-fifths were accounted for by manufactures. Since the Second World War, these proportions have been reversed. As Table 2.1 shows, the growth of trade in manufactures has consistently exceeded that of the two primary sectors by a considerable margin. As a proportion of world commodity trade, the share of manufactures rose from 35 per cent to 61 per cent in the third of a century covered by the table.

Of all the sectors, it is manufacturing which has experienced the greatest impact of the post-war trend to greater openness of national economies. This basic fact has three interrelated facets of particular relevance in the present context. First, it used to be the case that much of international commodity trade took the form known as inter-industry trade; a country might export primary products and import manufactures, or it might sell cars and buy steel. Since the Second World War, an increasing proportion of trade in manufactures has been intra-industry; manufacturing countries will both import and export cars, steel and a multitude of other commodities. This intra-industry trade arises from the very large amount of product

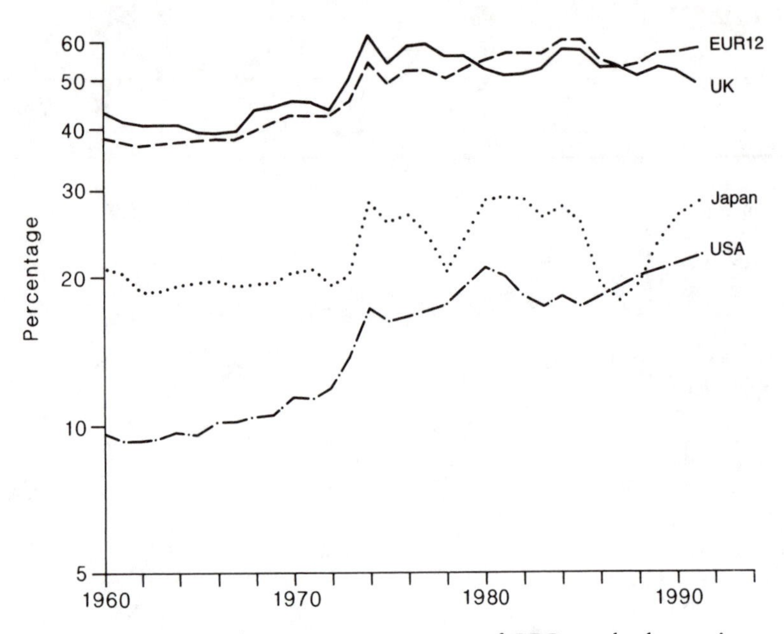

Figure 2.1a Imports plus exports as percentage of GDP, goods plus services

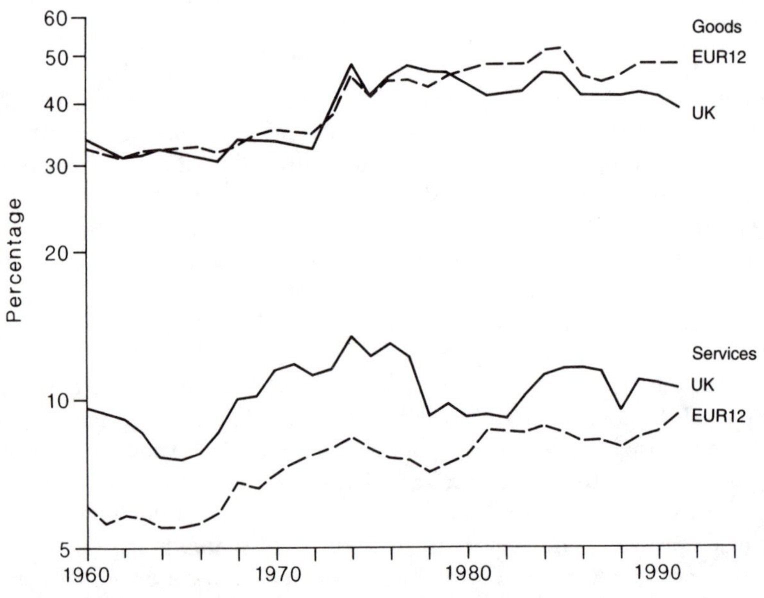

Figure 2.1b Imports plus exports as percentage of GDP, goods and services separately
Source: *European Economy* 46, 1991, Statistical Annex

Table 2.2 EUR12 countries, Japan and the United States: imports plus exports as a percentage of GDP, selected years

	Goods						*Goods and services*		
	Total			*Intra-EUR12*			*Total*		
	1960	*1972*	*1991*	*1960*	*1972*	*1991*	*1960*	*1972*	*1991*
Benelux	70.0	86.6	127.3	41.1	64.1	91.6	84.5	105.0	155.1
Denmark	58.0	43.3	54.1	32.3	20.1	28.5	65.6	53.6	67.0
France	22.8	26.9	41.2	8.4	16.0	26.1	26.9	32.4	47.3
Germany	31.5	33.5	53.8	12.6	17.8	29.5	35.0	39.6	62.7
Greece	a	25.5	40.6	12.2	14.1	26.2	a	31.7	53.0
Ireland	63.3	66.5	108.7	44.9	49.5	79.6	69.1	74.5	123.4
Italy	22.3	27.7	32.9	8.5	14.6	19.5	26.5	34.6	40.7
Netherlands	77.4	71.2	102.2	44.5	49.7	71.0	93.6	87.2	112.2
Portugal	37.1	41.4	69.9	17.3	20.8	49.2	41.2	59.1	83.7
Spain	12.9	19.5	29.1	6.1	8.8	17.7	17.6	29.0	39.1
United Kingdom	33.6	32.4	39.0	7.6	11.0	20.3	43.2	43.6	49.5
EUR12	32.3	34.7	48.0	12.7	19.0	28.9	38.4	42.5	57.3
USA							9.6	11.9	22.0
Japan							20.9	18.9	27.8

Notes: GDP measured at market prices; all figures in current prices

a The published sum of imports and exports of goods exceeds the equivalent figure for goods and services, implying an error of calculation

Source: *European Economy* 46, 1991, Statistical Annex

differentiation that has taken place and also from the expansion of trade in manufactured inputs – both the hallmark of modern manufacturing. For most of the mature industrialized economies, intra-trade now accounts for well over 50 per cent of trade in manufactures (Commerzbank 1992). The second facet, associated with the first, is the fact that trade between the industrialized countries is progressively becoming of ever-greater relative importance for these countries. The third feature of interest is that the growth in intra-trade implies that imports are taking an ever-larger share of domestic markets (import penetration has been increasing), while at the same time the proportion of output exported has been rising.

Figure 2.1 and Table 2.2 together summarize the impact of the trend to greater openness of national economies. Figure 2.1 shows the value of imports plus exports as a percentage of GDP for the EUR12 countries, for Britain separately, and for Japan and the United States. Figure 2.1a shows trade in goods plus services, whereas 2.1b distinguishes these two sectors, but only for the EUR12 countries and Britain. It is quite clear that since 1960 there has been a general tendency for the economies of the industrial nations to become considerably more open, but that this trend slowed quite sharply after 1974. It is also clear that although Britain's economy is about as open as is the case for the EUR12 countries collectively, there has been a divergence of trend since 1974, with Britain's economy becoming somewhat less open (see also Tsoukalis 1991: 196). Nevertheless, as Table 2.2 shows, over the period between 1960 and 1991, all the EUR12 countries except Denmark experienced an increase in commodity trade as a proportion of GDP. Of the four major economies – Britain, France, Germany and Italy – Britain has had the least growth in openness. Britain has for a long time been a trading nation, with a high level of imports and exports. Her economy was generally more open than the others in the early post-war period; since then, the economies of the other three countries have evolved in such a manner that they have been catching up with Britain or, in the case of Germany, actually going ahead.

TRADE BETWEEN THE INDUSTRIAL COUNTRIES

For the EUR12 countries as a whole, the increase in the openness of individual countries has arisen from the increasing importance of intra-trade among the group. Table 2.2 shows that between 1960 and 1991, total commodity trade as a share of GDP rose from 32 per cent to 48, i.e. by 16 percentage points. Over the same period, the significance of intra-trade also rose by 16 percentage points, from 13 to 29. Relative to trade with the rest of the world, commodity trade among the EUR12 countries has become considerably more important over the last three decades. Denmark is the clear exception to this general experience; in her case, the entire decline in openness on account of commodity trade occurred in respect of trade with

Table 2.3 Percentage distribution of total commodity exports by value, developed[1] countries and EUR12

	Exports from all developed countries destined for developed countries as % total exports from developed countries	*Exports from EUR12 as % total exports*	
		All developed countries	*Of which EUR12*
1953	59.6		
1955	60.3		
1960	66.0		
1965	70.5		
1970	73.8		
1975	68.3		
1980	69.4	77.4	55.8
1985	73.5	79.9	54.3
1990	77.2	83.7	60.5

Note: [1] Or industrial countries
Source: GATT 1992 and earlier editions. Comparable data are not available in later editions

the other EUR12 countries. Overall, therefore, there has been a substantial reorientation of commodity trade, to emphasize the growing relative importance of European connections. This reorientation has been most marked for Britain and the Netherlands, but especially in the case of Britain (Artis 1986; Curwen 1990). For both of these countries, commodity trade with the world outside the EUR12 group as a percentage of GDP, instead of remaining constant or rising slightly, actually declined. In other words, the realignment of trade patterns towards Europe was more marked for Britain and the Netherlands than for any other EUR12 country.

To what extent the reorientation of the trade of the EUR12 countries is attributable to the formation and enlargement of the EU is a matter to which we will turn in Chapter 4. For the present, though, we will note that Figure 2.1 and Table 2.2 suggest that the trends have occurred irrespective of the date a country joined the EU, implying that they form part of a general change in the structure of trade. Table 2.3 shows that over the period between 1953 and 1990, the developed (or industrialized) countries have been sending an ever-larger proportion of their exports to each other. From about 60 per cent, the share rose to 77 per cent in somewhat under four decades. Because there has been a change in definition, from industrialized country to developed country, there is an element of discontinuity in the series. However, this is relatively small, and the trend is both clear and strong. This trend is associated, of course, with the fact that trade in manufactures has been expanding much more rapidly than trade in other commodities. The table also shows data for the EUR12 countries in recent years. Overall, a bigger share of exports from the EUR12 countries goes to other developed countries than is true of these countries collectively, but the difference is not very great and, over the decade shown, has closed somewhat. This table also shows that there has been an increase in the proportion of EUR12 exports accounted for by intra-trade. That the intra-trade of the EUR12 countries amounts to three-fifths of their exports is hardly surprising, given that they account for over half the total exports of the developed countries (cf. Bhagwati 1991). Thus, although intra-trade among the EUR12 countries is clearly very important, it is essential to treat it as an element in the general situation that the more developed countries of the world have increasingly been trading with each other; on the world scale, the EUR12 countries are a major component of the developed world.

In sum, it is clear that the reorientation of trade among the EUR12 countries is to be interpreted as an element, and an important one at that, of the more general tendency for the industrialized nations to trade among themselves. While this cannot be the whole story, this basic fact is of central importance in any attempt to understand the changing relationship of Britain to the European community.

THE COMPETITIVENESS OF THE EU

In recent years, there has been some anxiety that the present EU of twelve countries may be losing its competitive edge in world markets, especially in the manufacturing sector (Francis and Tharakan 1989; Stopford 1993; Winters and Venables 1991). That there are grounds for concern is shown by changes in the share that the EUR12 countries have of world exports and imports of manufactured goods (Table 2.4). For the EUR12 countries collectively, the share of world exports of manufactures fell by almost two percentage points in the decade from 1980 to 1990, while the share of imports rose by just over three percentage points. These adverse trends are evident for Britain and also for the other eleven countries, as also for the United States and Japan, though Japan maintains a very large surplus in its trade in manufactured goods.

Thus, although the developed countries are to an increasing extent trading among themselves at the aggregate level, other countries are finding that they can out-compete them in many lines of manufactures. The tendency towards intra-trade among the developed countries is therefore to be regarded as the resultant of two opposing tendencies. First has been the relative decline of the primary sector in international trade, reducing the relative need for developed countries to import foodstuffs, fuels and raw materials. That process has operated throughout the post-war period. If nothing else had changed, the relative decline in trade in primary goods would have resulted in an increase in the proportion of trade (which is increasingly trade in manufactures) conducted among the developed countries. Second, though, in more recent years several developing countries, including Korea and Taiwan, have had a significant impact on trade in manufactures and are challenging to join

Table 2.4 Percentage shares of world trade in manufactures, 1980 and 1990

	1980		*1990*	
	Exports	*Imports*	*Exports*	*Imports*
EUR12 minus UK	38.6	29.9	38.0	32.7
United Kingdom	7.5	6.3	6.0	6.7
Total, EUR12	46.1	36.2	44.0	39.4
USA	13.3	11.0	11.9	14.7
Japan	11.2	2.2	11.3	3.9
Total	70.6	49.4	67.2	58.0

Source: GATT 1992

Table 2.5 Intra-trade in manufactures, advanced nations[1] and EUR12

	Intra-trade among the advanced nations as percentage of world trade in manufactures[2]	*Intra-EUR12 trade as percentage of total EUR12 trade in manufactures*[3]
1955	45.9	–
1963	54.1	52.1
1968	58.1	53.6
1970	59.6	56.0
1972	61.1	58.6
1973	60.0	58.9
1974	57.1	57.0
1975	53.8	55.7
1976	54.8	57.2
1985	58.5	55.6
1986	61.4	58.1
1987	61.6	59.4
1990	62.1	60.9
1991	–	61.0
1992	–	61.1

Notes: [1] Advanced nations defined as follows:
1955–76 USA, Canada, Japan, Western Europe (including Turkey and Yugoslavia)
1985– As above, plus Australia, New Zealand and South Africa
[2] *Source*: GATT 1992, vol. II, Table A.3 and earlier editions
[3] *Source*: *External Trade and Balance of Payments. Statistical Yearbook*, 1993, Eurostat

the club of developed countries (see, for example, Batchelor *et al.* 1980). The result of this challenge has been to limit the expansion of intra-trade in manufactures among the developed countries, as is evident if Table 2.5 is compared with Table 2.3.

Some further detail is provided by Figure 2.2, which shows, for the EUR12

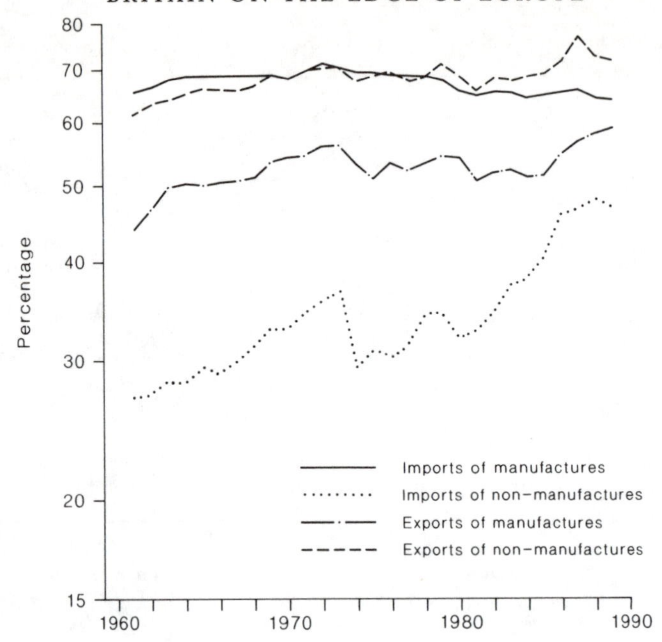

Figure 2.2a Intra-EUR12 trade as percentage of the total value of EUR12 international trade, EUR12 minus the UK

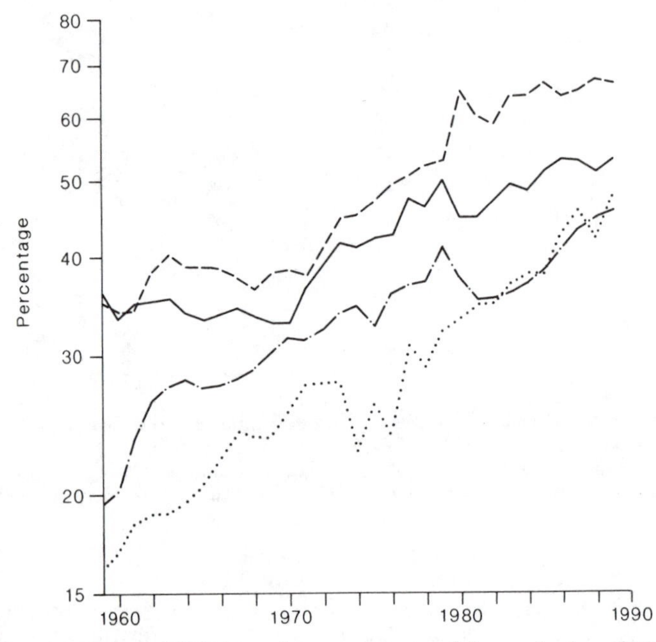

Figure 2.2b Intra-EUR12 trade as percentage of the total value of EUR12 international trade, UK

Source: *External Trade. Statistical Yearbook*, 1990, Eurostat

countries, the proportion of imports and exports derived from and destined to the other members of the group. Figure 2.2a summarizes the position for the eleven members other than Britain. For these countries, the share of manufactured imports derived from within the EUR12 group rose until 1972 but has since declined steadily. On the other hand, intra-trade has taken a steadily rising proportion of exports. The worrying implication is that manufactures originating in these eleven countries are becoming less competitive on world markets. Britain, shown in Figure 2.2b, has shared this experience in the sense that the proportion of manufactured imports derived from the EUR12 countries has risen more slowly than the share of exports destined to them.

Throughout the post-war period, the export of fuel and raw materials from the EUR12 countries to the rest of the world has been a minor component of Europe's trade (Table 2.6), as also has been the export of food, beverages and tobacco. The EUR12 countries collectively are manufacturing nations, so loss of competitiveness in this sector is a matter of considerable concern. On the other hand, primary imports have become steadily less significant, with the implication that an increasing share of imports competes directly with Europe's manufacturing base.

Table 2.6 EUR12 commodity trade with the rest of the world, per cent shares

	Imports			*Exports*		
	1958	*1980*	*1992*	*1958*	*1980*	*1992*
Western industrial countries	48	46	59	48	50	55
of which – EFTA	14	17	23	20	25	25
– USA	18	17	18	13	13	17
– Japan	1	5	11	1	2	5
Other countries	52	54	41	52	50	45
Products						
Fuel	16	35	13	8	7	5
Raw materials	30	11	7			
Food, beverages, tobacco	30	9	8	9	8	8
Machinery/transport equipment	6	14	30	37	37	41
Chemicals	17	24	37	9	10	12
Other manufactures	–	–	–	37	32	30
Miscellaneous	1	7	5	0	6	4

Sources: *External Trade. Statistical Yearbook*, 1990, p. 32, Eurostat; *External Trade and Balance of Payments. Statistical Yearbook*, 1993, Table 1 and Table 3, Eurostat

Taking the import and export evidence together, there seem to be justifiable grounds for concern that the manufacturing sector in the EUR12 countries is not as competitive as might be desired.

INTRA-EUR12 TRADE, WITH SPECIAL REFERENCE TO BRITAIN

Figure 2.2 confirms the point already made, that intra-trade among the EUR12 countries has been acquiring greater relative significance, except in respect of imports of manufactures into the eleven countries other than Britain. Of considerably greater interest in the present context is the fact that the reorientation of trade has been much more marked for Britain than for the other countries; for this country, the significance of intra-trade has converged to the level experienced by the others.

Country-by-country evidence confirms the fact that Britain's economy has experienced a more marked and more consistent reorientation towards the other EUR12 countries than any of the other nations. Figure 2.3 displays the position with respect to imports, and Figure 2.4 for exports. In both cases, the data relate to total commodity trade; also, Figures 2.3a and 2.4a give details for the original six members of the EU, while 2.3b and 2.4b deal with the countries that joined the EU after its formation in 1958. These figures also demonstrate that, for all the EUR12 countries, the significance of intra-trade has converged to the range 50–70 per cent of total imports and exports, but that the greater part of this convergence for the EUR6 countries occurred prior to 1972, most notably in the case of exports.

The reorientation of Britain's trade towards Europe has in fact been primarily with the six countries which formed the EU in 1958 (Table 2.7). These EUR6 countries accounted for little more than 10 per cent of Britain's trade in 1950 but over 40 per cent four decades later. Meantime, the share taken by the other five countries of the EUR12 group edged up only marginally, by two or three percentage points. All of the EUR6 countries have shown a strong increase in Britain's import and export shares, though the most striking increase has been with Germany (Table 2.8).

Figures 2.5 and 2.6 trace Britain's import and export share accounted for by the EUR6 countries, from before the Second World War to the present. From around 1950, there has been a remarkably steady increase in the proportion of trade conducted with these six countries, save only for part of the 1960s, when the export share seemed to be on a plateau.

It is quite clear that the shift in trading pattern has been a good deal more dramatic for Britain than for the other EUR12 countries, and that the reorientation has been remarkably steady over the post-war period. Taken in conjunction with the evidence previously discussed concerning changes in the global economy and the role of the developed countries therein, it is clear that some quite powerful forces are at work and that it will be necessary to

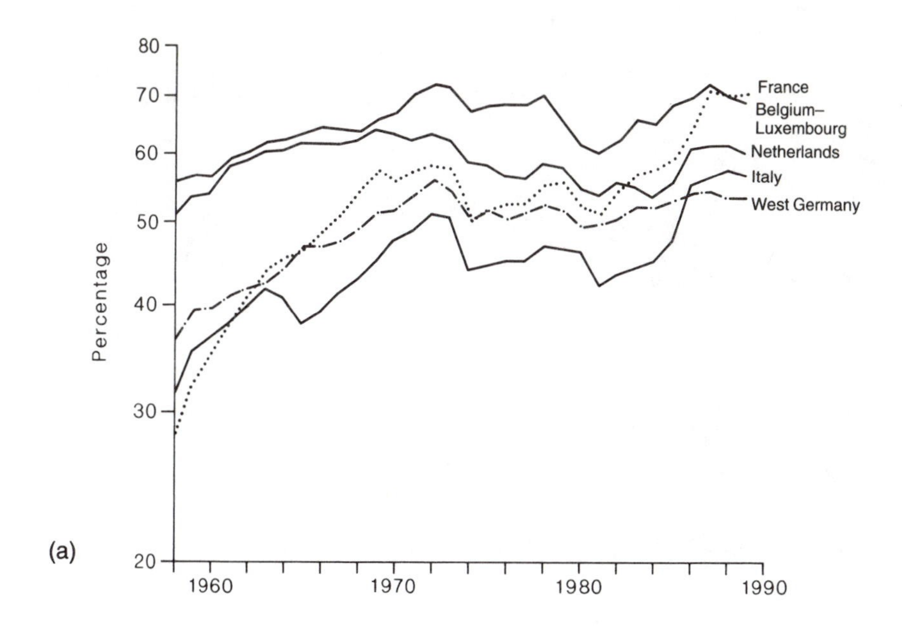

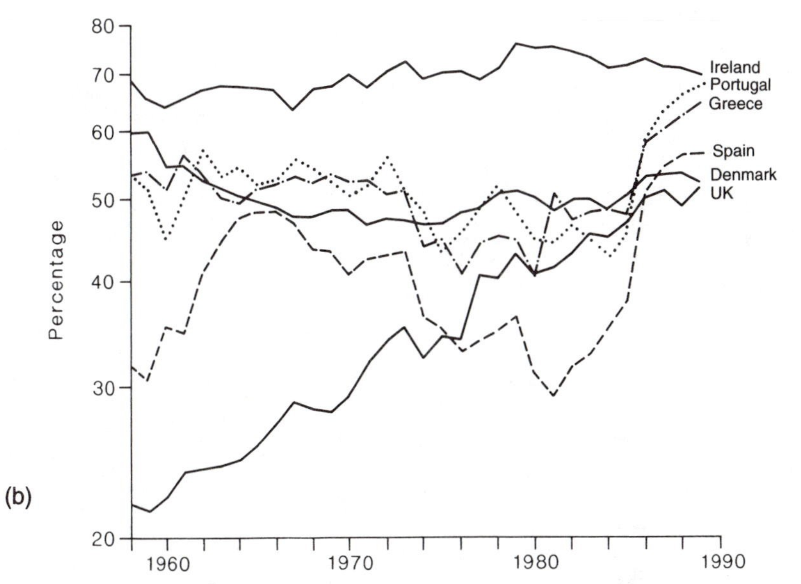

Figure 2.3 Intra-EUR12 imports as percentage of total value of imports, individual countries
Source: *External Trade. Statistical Yearbook*, 1990, Eurostat

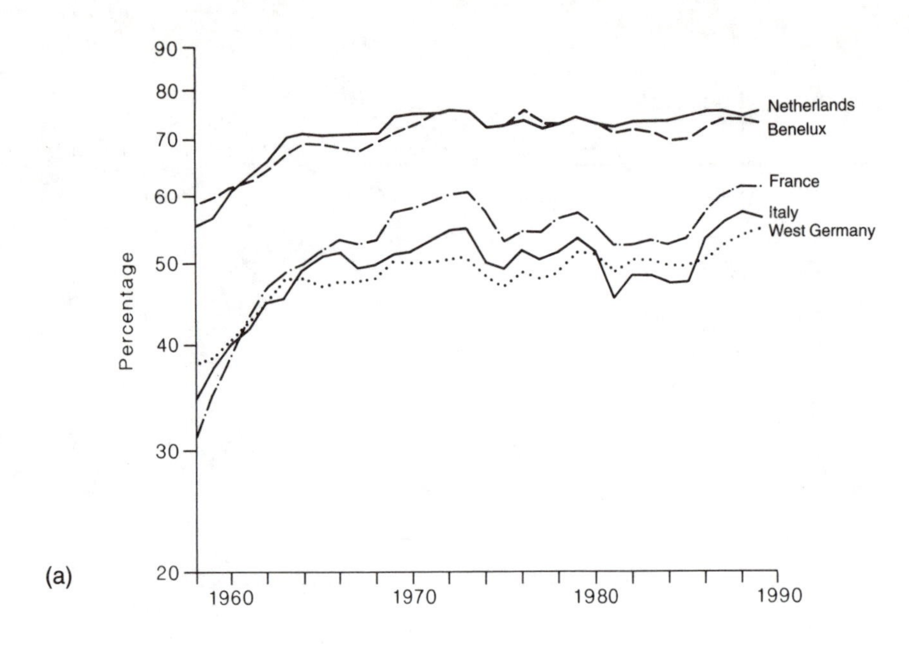

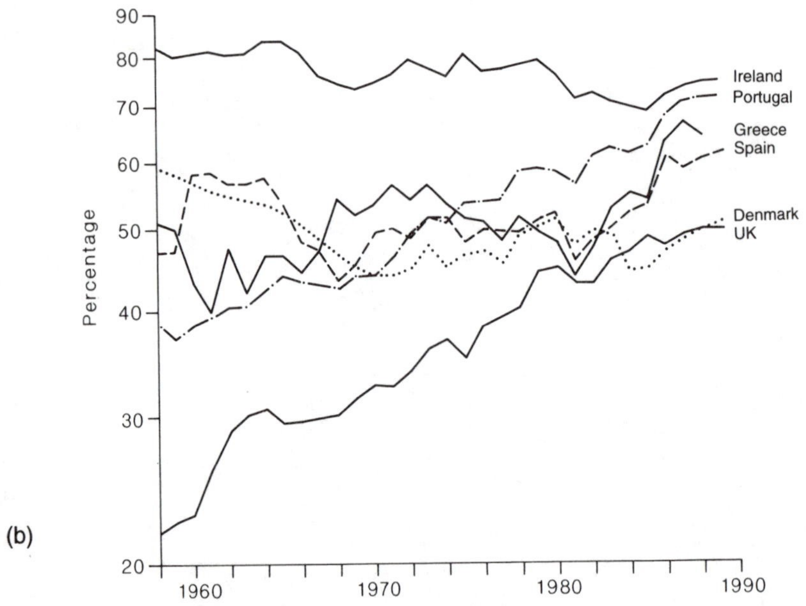

Figure 2.4 Intra-EUR12 exports as percentage of total value of exports, individual countries

Source: *External Trade. Statistical Yearbook*, 1990, Eurostat

Table 2.7 United Kingdom trade with the world, percentage distribution, all commodities by value

	1950	*1960*	*1970*	*1980*	*1990*	*1991*	*1992*
Imports							
EUR6	12.7	14.4	20.3	35.3	43.4	42.6	42.7
Rest of EUR12	7.8	7.9	9.5	8.4	8.8	9.0	9.5
Rest of Western Europe[1]	6.3	9.2	12.0	12.0	12.6	12.0	11.5
Rest of world	73.2	68.5	58.2	44.3	35.2	36.4	36.3
Exports							
EUR6	11.2	15.4	21.7	35.2	41.2	44.5	43.3
Rest of EUR12	8.8	7.6	11.0	10.6	11.8	12.1	12.6
Rest of Western Europe[1]	9.5	9.7	13.0	11.6	9.0	8.2	7.9
Rest of world	70.5	67.3	54.3	42.6	38.0	35.2	36.2

Note: [1] Mainly Austria, Finland, Norway, Sweden, Switzerland, Turkey, Yugoslavia
Source: *Annual Abstract of Statistics*

Table 2.8 Share of United Kingdom visible trade, EUR6 countries all commodities by value

	Mean percentage, three years			
	1949–51		*1990–2*	
Country	*Imports*	*Exports*	*Imports*	*Exports*
Belgium–Luxembourg	1.8	2.2	4.6	5.4
France	3.7	2.0	9.5	10.7
Germany	1.7	1.8	15.3	13.6
Italy	1.9	1.2	5.4	5.7
Netherlands	3.2	3.1	8.2	7.7

Source: *Annual Abstract of Statistics*

keep these clearly in mind in later chapters and, at least in part, to explore further. We should note, though, that the reorientation of Britain's trade towards Europe up until about 1970 was no more than a return to the position which had obtained prior to the First World War. Rowthorn and Wells (1987: 169) show that in 1913 the eight countries which, with Britain, constituted the EU after its first enlargement, accounted for about 30 per cent of Britain's trade but that this share stood at only 15 per cent in 1948; by the early 1970s, the proportion had climbed back to 30 per cent and by 1983 had reached 45 per cent. The fact that the geography of international trade can fluctuate so sharply suggests that we must exercise considerable caution in

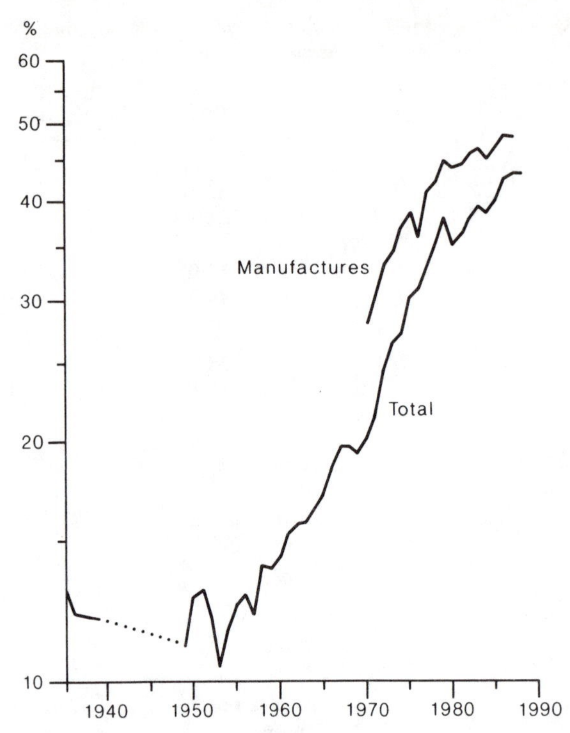

Figure 2.5 United Kingdom: imports from the EUR6 countries as percentage of total imports by value
Source: *Overseas Trade Statistics*

our approach to the question of what effect, if any, the location of Britain on the edge of Europe has on economic performance.

IMPORT PENETRATION: BRITAIN

The rising share of the domestic market taken by imports has been a matter of concern in some quarters (Blackaby 1978; Chisholm 1985a; Rowthorn and Wells 1987). However, to consider import penetration without at the same time examining the proportion of output which is exported is not very helpful, since in an open economy it is the relationship between the two which influences the balance of payments and other macro-economic variables. Figure 2.7 plots three sets of time-series data, constructed on somewhat different assumptions, showing the evolution of import penetration, export share and the ratio between the two for the manufacturing sector. The key facts are that since the 1950s there has been a steady increase in both import penetration and export share, but that the growth of the former has consistently outstripped the latter over the entire period. From the British

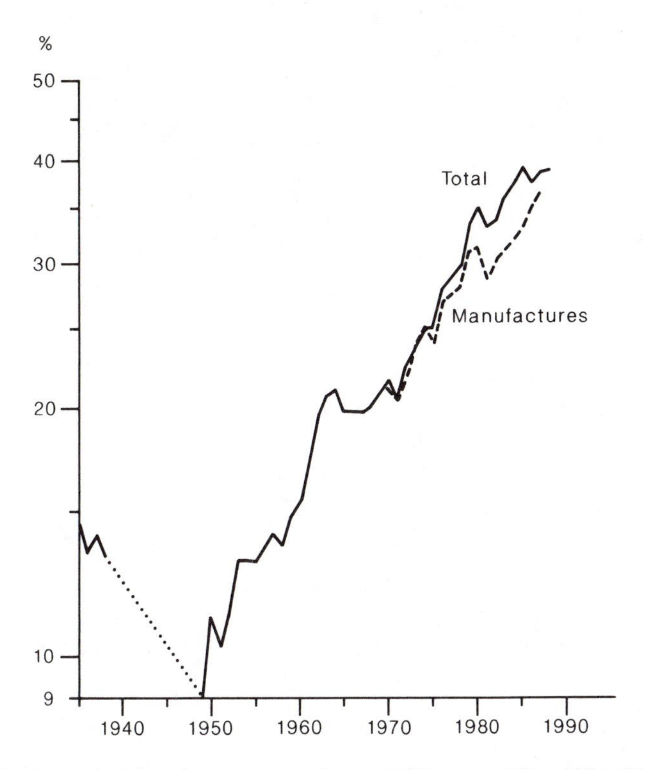

Figure 2.6 United Kingdom: exports to the EUR6 countries as percentage of total exports by value
Source: *Overseas Trade Statistics*

point of view, this long-term trend for export share to lag behind import penetration strongly suggests that British industry has become progressively less competitive.

Is there a geographical pattern to the shifting balance between imports and exports which is relevant in considering the relationship between Britain and Europe? In the absence of import penetration and export share data disaggregated by overseas geographical areas, a relatively crude means of answering this question is to take manufacturing exports as a percentage of manufacturing imports. These aggregate data are available for the whole post-war period, and from 1970 onwards it is possible to distinguish countries and groups of countries overseas. In the light of the previous discussion, showing the dramatic increase in the importance of the EUR6 countries in Britain's trade, this group of countries has been identified in Figure 2.8. Throughout the post-war period, the ratio of manufactured exports to imports has deteriorated. Since 1970, that deterioration has been evident in trade with the EUR6 countries and also with the rest of the world.

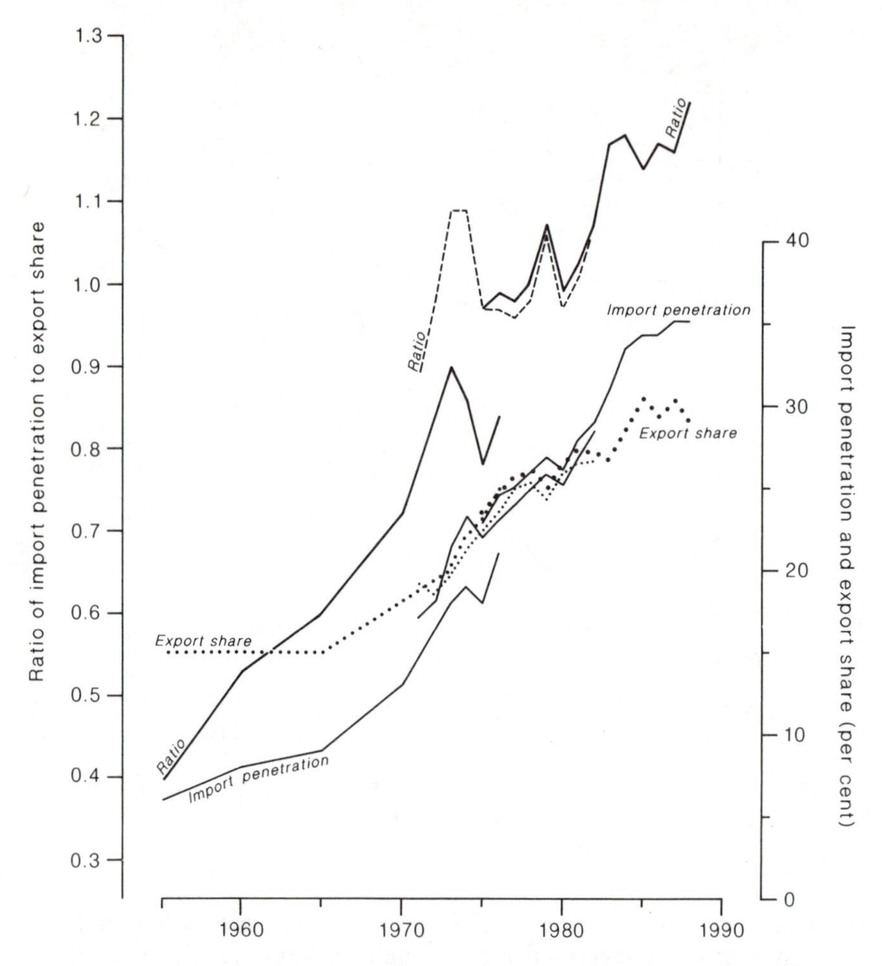

Figure 2.7 United Kingdom: import penetration and export share, manufactures
Sources: Brown and Sheriff 1978: 244; *Annual Abstract of Statistics*

The only difference between the two groups of countries is that the trade balance with the EUR6 countries was substantially less favourable in 1970 than it was with other countries, and has so remained; the rate of deterioration has been similar for both country groups.

The commodity structure of Britain's trade, especially of imports, has changed much more dramatically than has been the case for the other EUR12 countries, and this change has an important bearing on the calculation of import penetration. Figure 2.9 shows that since the early 1950s, manufactures have taken a steadily increasing share of British imports, rising from about 20 per cent to over 70 per cent. The two commodity classes which have lost

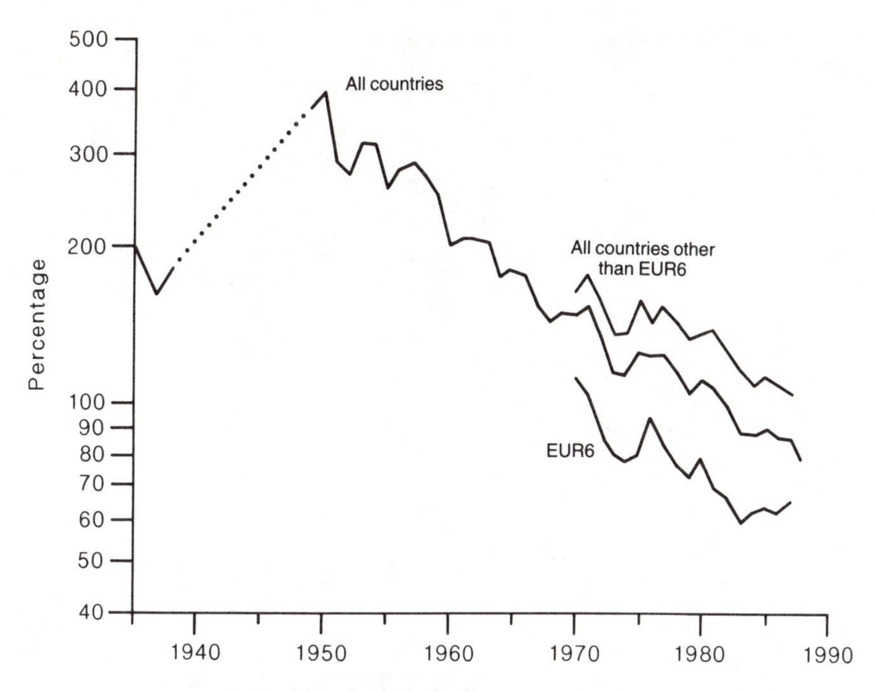

Figure 2.8 United Kingdom: exports of manufactures as percentage of imports of manufactures by value
Source: *Annual Abstract of Statistics*

share in nearly equal measure are foodstuffs and related products, and basic materials. The latter used to provide the main material input into manufacturing; the role of basic materials has been eroded by many factors, including the fact that an increasing share of physical inputs into manufacturing processes consists of partially manufactured goods (such as chemical feedstocks) and components. Comparison of Figure 2.9 with Table 2.6 shows that the transformation of the British economy with respect to the structure of imports has been much more dramatic than for EUR12 as a whole. On the other hand, if we ignore the temporary increase in fuel exports occasioned by North Sea oil, the structure of Britain's exports by broad sectors has remained very stable, as has been the case with the other European countries. Nevertheless, the overall structure of Britain's trade has changed substantially more than has been the case generally (Rowthorn and Wells 1987: Ch. 8).

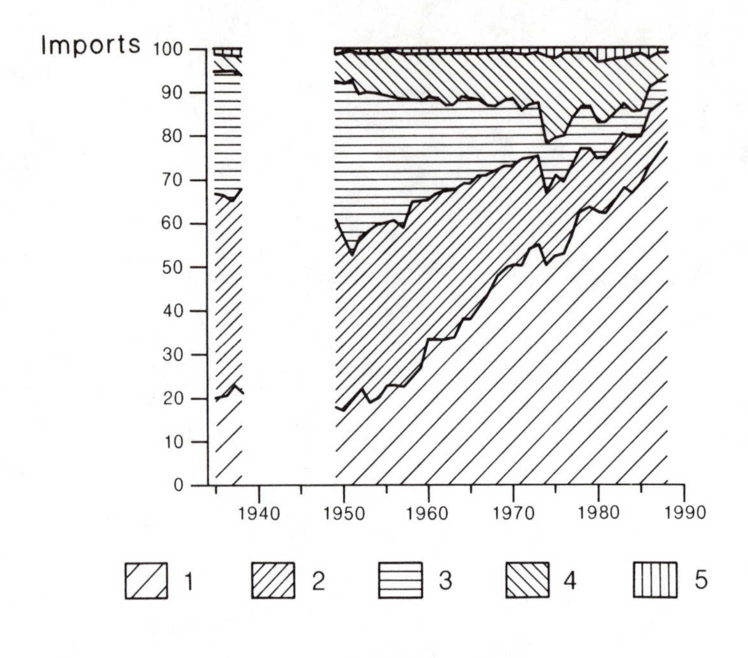

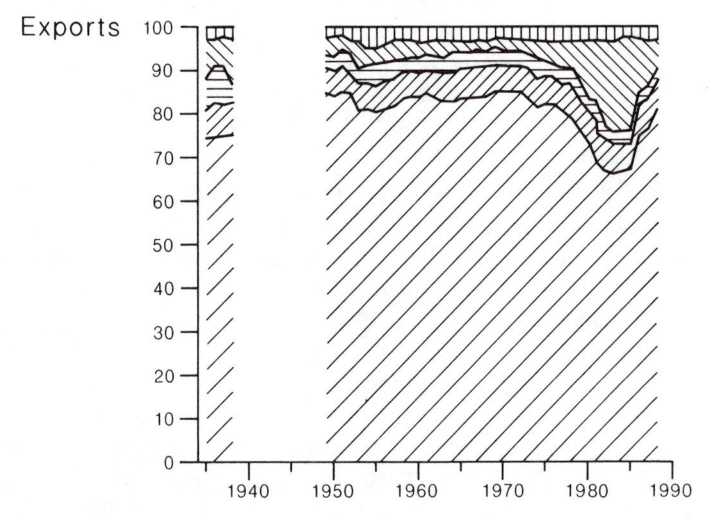

Figure 2.9 United Kingdom: commodity composition of foreign trade by value, percentage

1 Manufactured goods
2 Food, beverages and tobacco
3 Basic materials
4 Mineral fuels and lubricants
5 Other

Source: *Annual Abstract of Statistics*

THE EMERGENCE OF REGIONAL TRADE ASSOCIATIONS

The evolution of the world economy, and of the EU and Britain, has taken place under the watchful eye of GATT, an organization committed to fostering the freedom to trade without hindrance. Yet the EU itself is, in important ways, in breach of GATT rules, since the Union differentiates between the trade which countries conduct with other members of the Union and countries which are outside its jurisdiction. In the former case, tariffs have been abolished and, more recently, steps have been taken to eliminate other trade impediments. These benefits are not available to other countries. However, the other nations of the world, and GATT itself, have been willing to accept this dilution of the pure doctrine of international free trade because the EU has been a party to the general and worldwide reduction in trade barriers (especially tariffs) negotiated under the auspices of GATT.

However, the EU is not the only group of countries which have agreed to create free trade areas (Anderson and Blackhurst 1993; de Melo and Panagariya 1993). Until recently, the regional groupings have generally been associations of relatively small countries, and the success of these groupings has been rather less than that of the EU itself. Therefore, the existence and multiplication of these regional trade blocs had not seemed to pose a significant threat to the continuing liberalization of world trade. During 1993, things began to look otherwise. There was a real possibility that the painfully extended negotiations on the Uruguay Round of GATT might end in failure. Had that occurred, the fear was widespread that the nations of the world would retreat from the hard-won liberalization of trade and retire into defensive regional economic groups. These fears were given impetus by the agreement of Canada, Mexico and the United States to form their own North American Free Trade Area (NAFTA), which gained its crucial endorsement from the American legislature in November 1993. Had the GATT negotiations failed, there was a real prospect that the trade barriers would have gone up around NAFTA and the EU, and the rest of the world would have been driven to take equivalent action in self-defence.

Although these worst fears have not been realized, some observers take the view that the world is now entering an era in which regional trading takes priority over truly international trade, with 'the reassertion of economic geography' (*Economist*, 20 November 1993: 14). However, the key point to note is the following. Given that GATT has made giant strides in reducing trade restrictions, and given that its successor, the World Trade Organisation, will probably be able to maintain and consolidate these achievements, regional trade associations are much less significant than they were forty years ago, and therefore have a smaller impact on world trade patterns than used to be the case. If in practice trade patterns do become more regionalized, but do so within a liberal trading regime, it seems unlikely that any of the discussion earlier in this chapter would be invalidated. Indeed, there would be a reinforcement of the trend already noted for the European countries to become progressively more interdependent, with the further relative growth of intra-trade in the EU. But whether this has in the past been, or would in the future be, determined by the proximity of the trade partners, as argued by the *Economist*, or by other considerations is a matter that warrants explicit examination.

CONCLUSION

Although this chapter has covered a large field in rather limited detail, several facts are abundantly clear which are relevant for the discussion which is to be undertaken in the chapters which follow. In the first place, it clearly is impossible to consider Britain's relationship with Europe, and especially EUR12, without at the same time considering the place of Europe in the world economy. Self-evident this may be, but not everyone has accepted the logic of this truth. Second, the EUR12 countries have participated in the general tendency for the developed countries to trade with each other rather than with other countries. Consequently, the expansion of intra-EUR12 trade is to be seen, at least in part, as a symptom of this global trend rather than as a peculiarly European phenomenon. Third, it is clear that the above two points become more relevant and significant as national economies become increasingly more open, which fact also emphasizes the key importance that, fourth, attaches to remaining competitive.

None of these points directly accounts for the astonishing redirection of Britain's trading relationships towards Europe, but they do serve to suggest that this reorientation may differ in degree rather than in kind from the experience of other developed countries; however, one must note that until about 1970 this reorientation served only to restore matters to the situation which existed at the turn of the century. Finally, it is abundantly clear that many of the trends identified, and especially Britain's shift towards Europe, have proceeded over many years. While it may be an exaggeration to say that Britain joined the EU in a 'fit of despair', few commentators now deny that her accession did at least recognize the economic realities of the time and the then oncoming future.

All of this does leave unanswered a basic question of fact and interpretation. How far is the growth of intra-EUR12 trade, and of Britain's trade with Europe, a response to the advantages of proximity? For although there are clearly important worldwide trends affecting all the developed countries, intra-trade among the EUR12 countries is important and may be at least partly explained by the advantages of mutual proximity. If that should be the case, then Britain's geographical location could be, at least in some measure, a handicap. It is to this question that we turn in Chapter 3.

3

DISTANCE AND INTERNATIONAL TRADE

> Extreme locational disadvantage in the Far East Asian economies has not prevented spectacular advances in their market share in North America and Europe.
>
> (Delors 1989: 83)

The reorientation of Britain's trade towards Europe could be interpreted in either of two ways. If distance really does have a significant impact on international trade, one could interpret the rapid growth in Britain's trade with Europe as an overdue adjustment away from an artificial and disadvantageous geographical pattern based upon outmoded concepts of Commonwealth preference and other distorting factors. Such an interpretation would be consistent with continuing anxieties about the impact on Britain's economy of occupying a location peripheral to Europe, which is now our most important trade area. Such an interpretation would overlook the fact that the realignment of Britain's trade away from distant partners has been much more marked than the reorientation experienced by the EUR12 countries collectively and individually. If distance really is important, then Britain will have experienced a substantial gain relative to the EUR12 countries during the post-war period on account of changes in the geography of her trade.

An alternative interpretation of the post-war shifts in trade is that they have occurred for reasons unconnected with distance, implying that distance costs are small or even negligible. This interpretation would be consistent with the fact that in the early part of this century trade with Europe was much more important for Britain than was the case immediately after the Second World War. It would also fit with the increase in manufactured imports, manufactures having higher unit values than primary materials, with the implication that transfer costs will now be less significant than formerly. And it would also accord with the extraordinary rise of Japan, South Korea, Taiwan and other Asiatic countries in world trade.

To help in assessing these alternative interpretations we will examine first the theoretical framework which suggests that distance is an important factor in international trade, and then review the available empirical evidence to see whether the expectations derived from theory are supported in practice. Therefore, in this chapter we concentrate on the literature which explicitly addresses the role of distance in international trade.

THEORY

Classical location theory places considerable emphasis on the minimization of transport costs, or, more generally, the costs of engaging in transactions across space. For similar or identical commodities, consumers are assumed to be sensitive to differences in the prices charged by different suppliers (demand is price elastic). Furthermore, it is assumed that prices vary systematically in space, reflecting the costs of transferring goods from their point of production to the point of consumption. If everything else is equal, firms will be at an advantage if they are close to a large number of suppliers and consumers – the more so if there are substantial economies of scale in production (Isard and Peck 1954).

From these basic premises it is but a short step to the proposition that firms will supply nearby markets in preference to more distant ones. Such behaviour at the level of the firm should be evident in the aggregate trade pattern of regions and nations, interaction being strong between areas which are close to each other, and weak where trade partners are distant from each other. A common form of this general idea is the gravity model, in which the volume of interaction between two regions or countries is positively related to their sizes and inversely to the distance which separates them. 'Size' in this context is some measure of economic mass, such as total population, GNP or the value of external trade. 'Distance' is conceived to be economic distance, i.e., the cost of transport and other costs associated with transferring goods from one place to another. In practice, data problems often constrain researchers to use geographical distance as a proxy measurement for economic distance.

The formal development of these ideas, generally referred to as spatial interaction models, is available in numerous publications (e.g., Batten 1983; Batten and Boyce 1986; Chisholm and O'Sullivan 1973; Fotheringham and O'Kelly 1989; Haynes and Fotheringham 1984; Isard and Bramhall 1960). In theory, these models are applicable at all geographical scales, subject to it being reasonable to assume system closure. In practice, most of the literature is framed for the analysis of inter-regional flows either at the sub-national level or for a system of regions which exhausts the national territory. Applications to international trade have been comparatively few; for example, the Fotheringham and O'Kelly (1989) review concentrates entirely on intra-national problems. Two technical reasons may be advanced to explain the neglect of international trade:

1 Although good data are available on international trade, the use of nations as the geographical unit of study implies heroic assumptions about the measurement of distances between the countries, including the suitability of using a single centroid for a country as big as the United States.
2 To overcome this defect by disaggregating international trade into a matrix of regions trading nationally and internationally would be an immensely expensive undertaking. The nearest approximation, available for only a few countries, is the disaggregation of a single nation's trade by internal region of origin/destination, the external trade partners being whole countries or groups of countries.

Notwithstanding these practical difficulties, the logic of location theory and spatial interaction modelling does suggest that distance costs should have an influence on the geography of international trade. This has been argued forcibly, but only in theoretical terms, by Krugman (1980). Although there is evidence that distance (accessibility) does have some impact on the spatial pattern of intra-national transactions (Gordon 1976), there is also evidence that one must be cautious in assuming that distance is therefore of great significance in international trade (Peschel 1981). It is necessary, therefore, to examine the available empirical evidence to see whether trade patterns conform to theoretical expectations. First, we will review the evidence available for global trade patterns, and then consider intra-European trade.

GLOBAL TRADE PATTERNS

Linnemann's 1966 study remains the most comprehensive attempt to model international trade flows. The main variables that he used to explain the volume of trade between pairs of countries were the GNP of the exporter and of the importer and the distance (in nautical miles) which separates them. He used data for seventy-nine countries and obtained coefficients and levels of R^2 that compared with the earlier studies of Tinbergen and Pulliainen, both of whom employed data for smaller samples of countries (Linnemann 1966: 84). Linnemann himself calibrates four equations, all in logarithms to the base e, two using all trade flows and two using only those flows that were equal to or greater than $50,000 (nominally $100,000), in each case computing the equations with their nominal or 'real' GNP data. All four equations yield similar results, though the importance of the distance variable was marginally greater using all trade flows rather than just those over $50,000. His data refer to the mean values for the three years 1958, 1959 and 1960.

Unfortunately, Linnemann does not provide sufficient raw data to calculate standardized partial regression coefficients whereby to assess the relative importance of the GNP, distance and other variables (King 1969: 139–41). However, it is possible to gauge the significance of the distance variable, albeit in a much cruder way. For this purpose, we will use the parameters estimated using all trade flows and nominal GNP values (Linnemann 1966: 82, 84; from an equation which overall yields an R^2 of 0.79). These are:

GNP of export country	0.99
GNP of import country	0.85
Distance	–0.81
Constant	0.13

The above parameters allow one to calculate what the volume of trade should be for any pair of countries assuming different distances separate them. This shows the distance effect while GNP is held constant. By doing this calculation also for different sizes of country, the impact of country size can be measured holding distance constant. Table 3.1 presents the results. The United States had the largest GNP at $483 billion, with the United Kingdom second at $67 billion; Libya, with $0.09 billion, was the smallest economy in the study. If one compares the variation across the rows of Table 3.1 with the column variation, it is immediately apparent that country size contributes about one thousand times more variation to trade flow than does the distance variable.

These admittedly rather crude measurements from Linnemann's data indicate that the distance variable, although statistically significant, plays a limited role in the geographical patterns of international trade. This fact, which is not explicitly acknowledged by Linnemann, probably accounts for the difficulty which he had in making sense of the country-by-country variation in the distance exponent, noting, as he does, that: 'the individual standard errors of the estimates are so large that the differences [in the values of the distance exponent] are hardly significant' (Linnemann 1966: 92). As a

Table 3.1 Export volumes estimated from Linnemann's data (billion dollars)

		Volume of exports		
Exporter	*Importer*	*100 nautical miles*	*1,000 nautical miles*	*13,000 nautical miles*
USA	UK	368.5	18.4	2.3
UK	USA	279.5	14.0	1.8
USA	Small economy[1]	10.3	0.5	0.06
Small economy[1]	USA	4.4	0.2	0.03
Small economy[1]	Small economy[1]	0.002	0.0001	0.00001

Note: [1] GNP of $1 billion
Source: Chisholm 1995 (derived from Linnemann 1966, his equation AC1)

result, Linnemann approached the problem in a different way, by standardizing trade flows between pairs of countries on the basis of their respective trade volumes, to generate an alternative measure of the effects of distance on trade patterns. Summarizing these findings, he says:

> The results are very interesting, as they quantify the significance of a 'good' or 'bad' geographical location for the relative importance of foreign trade in the economy. At opposite ends of the scale we find (a) the Netherlands and Belgium, as countries with an 'ideal' foreign trade location, and (b) Japan, Australia and New Zealand as badly situated countries with long distances to the major markets. On the basis of our findings we are able to say that the effort involved in realizing a certain volume of trade is about six times greater for the latter countries than for those mentioned under (a). *Admittedly, the meaning of the word 'effort' in this context is somewhat vague; it should be understood as referring to the overcoming of the natural obstacles to trade.*
>
> (Linnemann 1966: 187; my italics)

This conclusion, while emphasizing the role of distance, fails to provide a useful measurement of the scale of the impact relative to other factors and in no way contradicts the conclusion we have already reached that Linnemann's evidence shows the impact of the distance variable in international trade to be rather small.

Linnemann (1969) returned to the problem of distance as a factor in international trade by first standardizing trade flows for the aggregate volume of imports and exports of trade partners and then taking the ratio of the expected volume of trade to the actual volume. He regressed these ratios, as the dependent variable, on the distance between pairwise trade partners. The regressions he obtained indicate that at a distance of between 1,500 and 2,500 kilometres the actual volume of trade would be twice the expected volume, whereas at distances under 30 kilometres there would be a hundredfold increase. These results are hard to reconcile with the 1966 study and also with further evidence from other authors, to which we now turn.

Johnston (1976) analysed the pattern of international trade for all the countries for which data were available, using 1960 and 1969. In fact, the figures for these two years are the mean values for the three years centred thereon. Two forms of analysis were used in an exploration of the effects of distance. In the first, Johnston derived the expected volume of trade on the assumption that trade between pairs of countries is proportional to their respective sizes as trading countries. The differences between these predicted flows and the flows actually observed were then regressed on distance. This is his 'indifference' model. In the second analysis, Johnston calibrated a gravity model, which he calls the total interaction model, similar to the model used by Linnemann, except that Johnston uses total imports or exports and population as the measures of 'mass', not GNP, and appears to pay more attention to the mass of the trade partners than to the mass of the country under consideration. As expected, the distance variable proves to have a negative impact on trade volume.

The problem is that in neither set of results does Johnston report the proportion of the trade flows accounted for by the various factors, and his data do not lend themselves to making this estimate. However, we can infer that the effect of distance must be very slight. Johnston's primary results are presented for each of the countries for which he had data – namely, the difference between the expected and the actual flows in his indifference model, and the distance coefficients in the gravity model. These are examined for the two years 1960 and 1969, and also for change between these years. The static analyses of these two variables yields inconclusive results. Of the gravity model distance exponents, for example, Johnston (1976: 83) says: 'The extremely poor results from these regression analyses suggest either that the spatial distribution of the ... coefficients is to a considerable degree random or that the underlying influences to the pattern were not being tested by the models employed here.' It is also apparent that, despite considerable stability in the overall trading system, there is little stability in the effect of distance:

> There is thus some stability, but not a great deal, in the strength of the distance influence according to [the indifference model].... Compared to the findings for the indifference model, those for the total interaction model indicate very little stability in the role of distance as a constraint to trade.
>
> (Johnston 1976: 122, 124)

As we have already seen, Linnemann also had great difficulty in making sense of the spatial pattern of the distance effect. Johnston's data confirm Linnemann's and also draw attention to the instability of the observed distance effect over time. All of this is consistent with the proposition that the greater part of the pattern of trade flows is accounted for by factors other than distance, primarily the size of the countries involved, as Johnston himself acknowledges (1976: 128).

Table 3.2 confirms this conclusion for the United States in the year 1987 in respect of exports of industrial products. The primary focus of the work of Erickson and Hayward (1991) is the geographical pattern of exports from the nine regions which comprise the contiguous United States. For this purpose, the authors compiled their own estimates of exports by region of origin and fitted a gravity model incorporating the factors shown in Table 3.2. They found significant collinearity between their variables for the difference in GDP between origin and destination and the volume of foreign direct investment in the United States, and therefore they fitted two versions of the model to the data. The information shown in Table 3.2 has been calculated from the standardized regression coefficients which they report

Table 3.2 United States, 1987: factors affecting the geographical distribution of industrial exports

	Percentage contribution to explanation of total variance	
	Regression A	*Regression B*
GDP of overseas country	36.6	27.1
GDP per caput difference between US and trade partners	–9.7	–
Foreign direct investment of trade partners in USA	–	24.7
Distance	–9.4	–8.5
European Community	–2.8	–6.1
COMECON	–13.3	–6.9
English speaking	4.9	2.1
Total (= $R^2 \times 100$)	76.7	75.4

Source: Erickson and Hayward 1991: 385

for the United States as a whole. The results show several things of interest. First, the single most important determinant of trade flows is indeed the size of America's trade partners measured by GDP. The existence of trade blocs and linguistic differences is the next most important factor, and distance accounts for between 8.5 and 9.4 per cent of the total variance. However, it will be noted that the estimated contribution of each variable is itself somewhat changeable, according to the precise model specification. In the present context, the key fact is that distance accounts for perhaps 9 per cent of the variance in export flows from the United States, a proportion which is somewhat higher than seems to be suggested by the more general work reported by Johnston and Linnemann but is nevertheless a comparatively small effect on trade flows.

The conclusion that distance is a relatively small factor in explaining global patterns of trade appears to be flatly contradicted by Hanink (1988). He develops some ideas previously proposed by Linder, who suggested that the growth of intra-trade among the rich countries could be ascribed to the similarity of their respective patterns of demand; goods produced and consumed in one country would find markets in others. Hanink proposes, therefore, that among rich countries the more similar the per caput income between two countries the larger the volume of trade. For poor countries, however, trade among themselves is not very rewarding, so they will both export to and import from the richer countries in disproportionate degree. On this reasoning, Hanink develops a regression model applicable to individual countries, in which the dependent variable is the value of imports drawn from the various trade partners divided by the population of the importer. Three independent variables are used to explain the pattern of imports per caput: the absolute difference in GNP per caput between trade partners; the absolute difference in population; and distance. This model, cast in natural logarithms, was applied to a system of twenty-six countries, yielding an average R^2 of 0.524, and showing distance to be much the most important of the three independent variables.

There are some fundamental problems with this formulation. In the first place, the Linder model applies to finished goods, and particularly to consumer goods. Yet a very substantial proportion of trade in manufactures is trade in semi-processed materials and components, i.e., manufactured inputs. In Britain's case, for example, 15 per cent of the 1992 export of manufactures was 'semi-manufactures' and the equivalent proportion for imports was 19 per cent. These figures understate the true position, in that fully completed components and parts are not treated as semi-manufactures. Clearly, there must be some doubt about the formulation of a model based on Linder's work. Second, and more important, it is clear that the size of trade partners is an important factor determining trade flows, and given that the gravity model is formulated in logarithms, the relationship between trade volume and country size is not linear in natural numbers. Yet Hanink removes country size in respect of both importer and exporter. Given that his overall level of R^2 is considerably lower than the 0.75 or 0.8 that has been obtained by other workers, his results cannot be said to be an improvement. In any case, Linnemann included both GNP and population as variables, which implicitly introduces differences in GNP per caput, while Erickson and Hayward incorporate this explicitly, albeit as the difference in GDP rather than GNP.

Hanink's conclusion that distance is the single most important factor determining the geography of trade per person does not look secure in the light of the evidence already reviewed. It looks even less secure in the light of other evidence to which we will now turn, and in particular the structure of ocean freight rates. This evidence shows that transfer costs are only rather loosely related to the distance over which goods are shipped internationally. Lipsey and Weiss (1974) report the most thorough extant study of the structure of ocean freight charges, using data for imports to the United States for seventy-six SITC commodities which cover the range from raw materials to manufactures. They take as their dependent variable the import freight charged in US dollars per metric ton of cargo. The three main independent variables for which data were obtained were the unit value of the commodities, the volume of space occupied by one metric ton of each commodity and the distance over which consignments travelled. Thus, the raw data are average figures for each commodity. The equations are calculated in logarithms, and although this is not specified in the text, these are to the base e and not to base 10.

Table 3.3 reports an exercise similar to that performed on the Linnemann

Table 3.3 Ocean transport charges estimated from the data of Lipsey and Weiss ($US per metric ton)

Maximum UV[1] and ST[2]		*Minimum UV[1] and ST[2]*	
Maximum distance[3]	Minimum distance[3]	Maximum distance[3]	Minimum distance[3]
649	374	2.1	1.2

Notes: [1] UV is the unit value, $ per metric ton: maximum 22,000, minimum 8
[2] ST is the stowage factor, cubic feet per metric ton: maximum 206, minimum 9
[3] Distance in nautical miles: maximum 9,765, minimum 1,906
Source: Chisholm 1995 (derived from Lipsey and Weiss 1974, their equation 15)

data, taking the reported maximum and minimum figures. It is immediately apparent that the recorded freight charge is not quite doubled if the distance of shipment increases from 1,906 nautical miles to 9,765 (these are the minimum and maximum values reported). However, the nature of the consignment – its unit value and the number of cubic feet per metric ton – has a far greater impact on the freight cost, the ratio of lowest to highest charge being just over 1:300. The relative unimportance of distance is confirmed by the standardized partial regression coefficients. These are: unit value 0.817, stowage factor 0.216, distance 0.090, derived from an equation which yields an overall R^2 value of 0.8. Hence, of the 80 per cent of the variance in freight rates accounted for by the regression, almost three-quarters is attributable to the unit value of consignments, just under one-fifth to the stowage factor, and only 8 per cent to distance. The implication is that distance accounts for only 6.4 per cent of the total variance in freight rates (8×0.8).

This evidence is consistent with an earlier study by the Organization for Economic Co-operation and Development (OECD) (1968). The sample consisted of consignments of manufactured goods in the trade between Europe and the United States in the mid-1960s. Taking the total cost from inland origin to inland destination, the ocean transport cost amounted to between 55 and 60 per cent, i.e., less than two-thirds of the overall transfer cost. The ocean transport cost included the loading and discharge costs, and other costs associated with port usage. These terminal costs amounted to somewhere between 40 and 50 per cent of the ocean freight element. It follows that, on average, only between 22 and 30 per cent of the total origin–destination cost in attributable to that element of ocean shipping cost which is related to the distance shipped. For individual commodities, the share of sea freight in total transport costs from inland origin to inland destination ranged from a low of 8 per cent in the case of automobile spare parts to a high of 84 per cent for packaging materials (Table 3.4). In other words, as Lipsey and Weiss found, the two factors which dominate freight costs are unit value

Table 3.4 Trade between Europe and the United States: breakdown of total transfer costs from inland origin to inland destination

	Automobile spare parts (%)	*Packaging materials (%)*
Sea freight[1]	8	84
Inland freight	59	7
Port charges, dues and other[2]	33	9
Total	100	100

Notes: [1] Sea freight includes the bulk of port costs, on average
[2] Direct charges only
Source: OECD 1968: 31

and the volume of space required per metric ton, so it is little wonder that the distance variable has a small impact on the volume of international trade.

An examination of Canadian exports throws some further light on the role of ocean transport costs. In this case, Bryan (1974) posed the following question: to what extent does the cost of liner freights affect the level of demand for Canadian exports in various markets? Data were compiled for twenty-five commodities, shipped to twenty-two countries which have direct access to oceanic traffic, the freight rate being calculated as a percentage of the f.o.b. price. Multiple regression techniques were used to estimate gravity models for each commodity. The results obtained showed quite clearly that demand for Canadian goods was affected by ocean freight rates only in the case of raw materials and bulky semi-processed goods:

> If these elasticities are correct, it can be inferred that Canada encounters particularly strong price competition in the overseas markets for lumber and newsprint. The fact that transport costs were not a significant explanatory variable for commodities such as wire and cable, copper tubing, construction machinery, aircraft engines and assembly, agricultural machinery, telephone apparatus, and card punching machinery probably suggests that factors like design, quality, tastes, guarantees, speed of delivery, and service conditions, are more important for these and similar commodities. The results may indicate that transport costs are more important for primary manufactures than secondary manufactures, the main reason being product differentiation.
>
> (Bryan 1974: 651)

These Canadian findings have been confirmed by evidence for the export performance of the United States (Kravis and Lipsey 1971). For the products covered in this study, average outbound freights were probably somewhat in excess of 10 per cent of the wholesale prices of the goods shipped. Yet

Table 3.5 Relative importance of factors explaining US export success

Factors underlying ability to export	*Crude materials (SITC 2) (%)*	*Chemicals (SITC 5) (%)*	*Manufactures classified by material (SITC 6) (%)*	*Machinery and transport equipment (SITC 7) (%)*	*All products (%)*
Prices equal to or below foreign	43	56	18	14	28
Product more expensive but off-setting characteristics	42	30	66	70	57
Unique product: no close foreign substitute	6	5	10	14	10
Other	10	8	6	3	5
Total	100	100	100	100	100

Source: Kravis and Lipsey 1971: 153

evidence from a sample of twenty-six firms, which together accounted for about 2 per cent of American exports, showed that delivered price was not a very significant factor in competitiveness in overseas markets, except in the case of crude materials and chemical products (Table 3.5). For most manufactures, non-price considerations dominated. A comparison of American exports of factory equipment with German imports of the same category showed that in only 7 per cent of cases was price the determining factor; product differentiation and service considerations accounted for 93 per cent of decisions.

These results are consistent with more qualitative data for the United Kingdom relating to the choice made by domestic firms between suppliers located in the United Kingdom and abroad (Kravis and Lipsey 1971: 154–7) and more general measures of the competitiveness of British manufacturers (Chisholm 1985a). They are also consistent with more recent work that has attempted to measure the sensitivity of EU trade to price changes, a matter of considerable interest in estimating the probable impact of the SEM on trade patterns (see Chapter 4):

> Thus all the evidence we have collected so far is pushing us towards the conclusion that international trade flows are less price sensitive than has been presumed over recent years – in particular, less sensitive than assumed in most exercises on '1992'.... Lower estimates of price elasticities mean that substantial changes in relative prices would be necessary to generate significant changes in trade shares.
>
> (Brenton and Winters 1992: 279, 280)

These findings carry a clear implication. If price, including the costs of transport, plays a rather limited role in the competitiveness of products in world markets, it follows that spatial variations in international transport costs will have a small, even negligible, impact on trade patterns.

Finally, Balassa (1986) has examined the determinants of intra-industry trade in manufactures in the overseas transactions of the United States, using 167 industry categories traded with thirty-eight countries. He hypothesized, *inter alia*, that the significance of intra-industry trade should be negatively associated with transport costs. Although the coefficient for the transport cost variable has the correct sign, it has a very low statistical significance, falling below the 10 per cent level when zero observations for intra-industry trade are included. The weakness of the association with distance costs serves to confirm the findings already reported.

Two main conclusions may be drawn from the evidence cited above. First, the effect of distance on international trade appears to be much less than theoretical considerations would lead one to expect. Although the distance coefficient is statistically significant and has the correct negative sign, the contribution which this factor makes to the volume of trade between pairs of countries is small. To the extent that distance is significant, Linnemann found that Western Europe, and especially Belgium and the Netherlands, occupies the best location for international trade; at the global scale, Britain shares the advantage of this good location and any difference from our neighbours on the mainland is too small to be measured with any confidence. Second, an important reason for the apparently small impact of distance on trade volumes may be found in the somewhat limited role of price in the competitiveness of manufactured products and the small impact which distance has on ocean freight rates. It appears that non-price characteristics are important, even dominant, in determining the competitiveness of products. If price variations arising from international transport are small, the impact of distance in trade volumes will be limited. Although international transfer costs add about 10 per cent on average to f.o.b. prices (Finger and Yeats 1976; Kravis and Lipsey 1971), distance seems to play a rather small part in variations about that average. Taking these two considerations together, it seems clear that, at least for major industrial countries, relative location within the global trading system does not greatly matter.

INTRA-EUROPEAN TRADE PATTERNS

The best-known study of the spatial structure of intra-European trade was published in 1956 by Beckerman. Unfortunately, he structured his analysis by standardizing trade flows for the size of the countries involved in pairwise flows and does not report the proportion of total trade which is thus accounted for. His spatial analysis is the analysis of residuals whose significance cannot be gauged. There are similar problems with other, more recent, studies (for example, Brams 1966; Peschel 1985; Prewo 1974). But Yeates (1968) reports the results of a gravity-model study of Italy's trade, with the impact of distance fully incorporated. His model includes cardinal data for the national income of the trade partner, its per caput income and the distance from Italy, as well as five dummy variables. Distance was taken as the square root of the distance along the arc of the great circle route connecting Rome with the capital of the trade partner. Equations were calculated for imports and exports in 1954 and 1965, and standardized partial regression coefficients are presented for each of the four equations. These standardized coefficients show considerable variation between the import and export equations and also between the two years, which raises questions concerning the robustness of the findings. Of more immediate interest, though, is the relative importance of the distance variable. The standardized partial regression coefficients for distance account for between 2.8 and 9.4 per cent of the variance in the trade flows, the mean value for the four equations being 5.6 per cent.

A small number of more recent studies has been published, using the gravity-model approach and employing data for regions within some of the nations of Europe (Bröcker 1988; Bröcker and Peschel 1988; Peschel 1981). In part, the focus of these studies has been the geographical impact of economic integration in Europe. However, by considering the distance effect explicitly, and by treating tariff changes as changes in distance, some light is thrown on the problem which we are considering. However, to construct inter-regional trade flows requires the use of rather heroic assumptions, so that the results must be treated with considerable caution. Bröcker (1988) concludes that most of the spatial variation in integration effects that he could detect arose from the sectoral composition of production in each region and not from the impact of relative location. The basic reason for this lies in the low values of the distance coefficients which he obtained. Peschel's review of the available evidence, mainly for Europe, led her to conclude:

> With respect to international trade the impact of distance is of minor importance compared to the influence of cultural and linguistic affinity of spatial association and political relationship.
>
> (Peschel 1981: 605)

We can explore these matters somewhat further with Eurostat data on the value of trade between the EUR12 countries. To examine these data, a fundamental choice must be made between the use of a gravity model to obtain estimates of the distance effect directly, or the standardization of flows for the importance of trade partners and the examination of the residuals to see whether their values are linked to the distance variable. Two sets of reasons suggest that the latter approach should be used.

The studies which we have already reviewed suggest very clearly that the primary factor which determines the volume of trade between pairs of countries is their respective size, measured by GNP or trade volume. There clearly is a logic, therefore, in using an approach which standardizes for this factor, but which does so in a manner which allows one to ascertain the proportion of total trade which is thus accounted for.

Mention has been made of the problem of measuring the distance between countries; it is time to explain the nature of the problem. The most commonly used measure of the distance between countries is based on the selection of centroids. Thereby, it is assumed that all trade originates and terminates at the same spot. Distance is then the distance between these centroids, measured by great circle route or along the routes of commerce used – railway, road or seaway. If a country is as small as Luxembourg, the assignment of all origins and destinations to one point does not do great violence to the reality. For a larger country, such as France, Germany or Italy, the use of centroids introduces a large amount of specification error, especially if economic activity is unequally spread (which it usually is) and if the country's shape diverges greatly from that of a circle (which it usually does). If the countries are far removed from each other the specification errors may be reasonably small, but where they are close together, as in

Europe, the errors are likely to be gross.

An alternative approach would be to obtain direct evidence on the cost of shipments. Given the multiplicity of routes and modes used for intra-European trade it is not possible to adopt the approach employed by Bryan (1974) in her study of Canadian exports, in which she obtained freight rates to different destinations. However, less direct information can be compiled to show the total transfer costs of trade between countries by comparing the f.o.b. values for commodities in the export statistics of countries and the matching c.i.f. statistics for the same goods recorded as imports. The difference should represent the sum of transport, insurance and other transfer costs. However, there are major technical problems in this approach (Beckerman 1956; Chisholm 1959; Geraci and Prewo 1977) and, in addition, a conceptual objection which is fatal. Comparison of f.o.b. and c.i.f. statistics provides information relating to the actual volume of trade, and this volume is clearly affected in some degree by transfer costs. Potential trade which does not take place because transaction costs are too high is excluded. In addition, an average cost obtained for several commodities will understate the impact of distance, since flows which are much affected by distance costs will be small or negligible and will play a small part in the overall figure. It was this fundamental objection which led Linnemann (1966) to reject the use of f.o.b. and c.i.f. data to measure the economic distance between countries.

In considering the problems of centroids and the difference between f.o.b. and c.i.f. prices, Beckerman (1956) concluded that, in the study of intra-European trade using national data, it is possible only to rank-order countries on an ordinal scale of proximity. This must be the correct conclusion, but it means that a reliable gravity-model approach is out of the question, unless national data are disaggregated into an inter-regional matrix.

With these considerations in mind, it seems evident that a gravity model cannot be meaningfully applied to intra-European trade analysed in terms of national units. Therefore, one is driven to consider an examination which starts by scrutinizing trade flows between pairs of countries according to some measure of their respective sizes. According to Linnemann's work, this should account for a large proportion of the volume of observed trade; the residual flows can then be examined to see whether nearby countries have a larger volume of trade than expected, whereas distant partners should have a lower volume.

The European Commission has published data on the bilateral trade flows between the EUR12 countries in 1958 and 1989 (Belgium and Luxembourg are combined, making an 11×11 matrix). These data have been analysed by Altham *et al.* (forthcoming). For any pair of countries, the bilateral volume of trade should be proportional to the aggregate volume of exports from the exporting country and the aggregate imports of the importing country. The larger these aggregates are, the larger the bilateral flows should be. If country size, measured by aggregate trade, were the only factor which determined

Table 3.6 The impact of country size and distance upon aggregate intra-EUR12 trade by value, 1958 and 1989, values of R^2

	1958		*1989*	
	Imports	*Exports*	*Imports*	*Exports*
Trade volume of importers and exporters alone	0.82	0.86	0.92	0.91
Trade volume of importers and exporters plus distance	0.83	0.88	0.95	0.93

Source: Altham *et al.* forthcoming

bilateral trade flows, then the estimated bilateral flows would exactly match the observed flows. However, to test whether distance is also a significant explanatory variable, a distance term was also included in the analysis, distance being measured as the road distance in kilometres between the capital cities.

Table 3.6 presents the results of this analysis, from which the following conclusions can be drawn. The aggregate trade volume of partners accounts for between 82 and 92 per cent of the variance in bilateral trade flows; the addition of a distance variable adds between 1 and 3 per cent to the explanatory power of the gravity model fitted. Clearly, within the European domain, the size of trade partner is far and away the dominant factor in explaining the geography of trade, and the contribution of distance is negligible, although there is a measure of collinearity between distance and country size as measured by trade volume.

Consideration was given to the residuals from a model based solely on the size of the trade partner. For each of the four sets of data represented in Table 3.6 (imports and exports for each of the two years), there are 110 flows, making an aggregate of 440 when the data are pooled. In the case of only twenty-two flows are the residuals significant at the 95 per cent level (two-tailed test). Of these twenty-two residuals, seventeen are positive, signifying that the observed bilateral flow is larger than expected. Almost all of these positive residuals are for trade between countries which are contiguous in the sense that they share a land border or are immediate neighbours, albeit separated by a short sea journey. Note also that of these seventeen residuals, eight are for trade between Ireland and the United Kingdom. The five negative residuals are for trade between widely separated countries, but three of these five residuals involve the trade of Ireland. There is, therefore, some evidence that distance does have an impact on trade flows, but the impact is slight. In so far as any one country is particularly affected, this clearly is Ireland; twelve out of the twenty-two residuals relate to this country's trade.

These results provide clear evidence that within Europe the impact of distance on the volume of international trade is very small indeed. They confirm the conclusions reached earlier in this chapter concerning the limited significance of distance in determining patterns of international trade.

CONCLUSION

The formal logic of locational analysis and of international trade theory suggests that the transfer of goods from the point of origin to another point of consumption will necessarily involve an increase in price. If this increase is not offset by scale economies in production, the price elasticity of consumption will ensure that sales will diminish with the distance shipped, if everything else is equal. Several important works published in the post-war period appear to confirm this expectation. However, when these are re-examined it turns out that the evidence for the distance effect is much weaker than had been supposed. This reinterpretation is confirmed by a new analysis of the trade of the EUR12 countries in 1958 and 1989: the effect of distance proves to be negligible.

This apparently surprising result can be explained by the fact that transport costs are low relative to the value of goods and that in any case a large portion of transport costs in international trade is invariant with the distance over which goods are shipped. In addition, the competitiveness of goods on international markets is only partially determined by price. For many goods, especially engineering and electronic products, vehicles, electrical goods, etc., the non-price characteristics of design, reliability and availability are more important than small price differences. In addition, as McConnell (1986) has argued, international trade patterns are increasingly shaped by the geography of foreign direct investment, the determinants of which are complex but manifestly more governed by productivity differentials, trade rules, rules about the sourcing of components, etc., than by distance costs.

The significance of this conclusion may be illustrated by two examples, both of which involve relatively bulky commodities for which one would expect transport costs to be an important factor shaping international trade. To protect the coastal cliffs near Looe in Cornwall from the erosive effects of high seas, a large quantity of stone in blocks weighing up to five tonnes each has recently been brought in from Norway, for the reason *inter alia* that it was cheaper to do so than use local stone (*Independent*, 13 May 1992). To ship a bottle of wine from Adelaide or Sydney all the way round the world to Britain costs about 10p. Not only is that well below 5 per cent of the retail value of most wines, it is also 'almost what it would cost to ship [a bottle] from some of the more inaccessible regions of France and Italy' (*Sunday Telegraph*, 16 January 1994).

In the light of the evidence presented in this chapter, it is clear that Britain's geographical position on the edge of Europe is a matter of little significance for trade and hence for prosperity. However important the country's position may be in political and cultural terms, in economic terms its relative location is unimportant. If Japan, for example, can overcome the 'extreme locational disadvantage' of its Asiatic position, the existence of 35 kilometres of water between Britain and France cannot be regarded as a serious handicap for British producers.

4

CORE AND PERIPHERY

Economic integration

> In reality, almost all existing cases of economic integration were either proposed or formed for political reasons even though the arguments popularly put forward in their favour were expressed in terms of possible economic gains. However, no matter what the motives for economic integration are, it is still necessary to analyse the economic implications of such geographically discriminatory groupings.
>
> (El-Agraa 1990a: 79)

There are good theoretical arguments for the proposition that economic integration between a group of nations will increase their aggregate wealth and, at the same time, effect some geographical redistribution of that wealth. This redistribution may occur at two geographical scales relevant to the present discussion: it may affect the relative prosperity of the countries involved; and it may trigger shifts in the relative standing of regions within one or more countries. The theoretical arguments are persuasive but, as we shall see, the empirical measurement of unification effects is a task which is fraught with considerable difficulty.

The post-war period has seen the publication of a very large literature on the economics of integration (e.g., El-Agraa 1989; Balassa 1962, 1989; Greenaway *et al.* 1989; Jacquemin and Sapir 1989; Jovanovic 1992; Molle 1990; Nevin 1990; Robson 1980; Scitovsky 1958; Viner 1950). By 1962, Balassa felt the need 'to present a unified theory of economic integration that would include, over and above the received theory, the dynamic aspects of economic integration, and would bring together the theoretical problems involved in co-ordinating economic policies in a union' (Balassa 1962: ix). For Balassa, the term 'economic integration' has two meanings: one as a process and the other as a state of affairs. As a process, economic integration signifies the reduction and ultimate elimination of discrimination between enterprises located in different nations. As a state of affairs, economic integration means the absence of various forms of discrimination which may occur between nations, a level playing field on which all economic actors are placed on an equal competitive footing.

Balassa recognizes several forms of economic integration. The simplest is a free trade area, in which tariffs and quantitative restrictions on trade between the participating countries are abolished, but each member retains its own tariffs against countries outside the group. The European Free Trade Association (EFTA), established on 1 July 1960, is an example. A customs union goes one step further, by setting a common external tariff, so that there is no incentive to route imports via low-tariff countries. It was this form of integration which the EUR6 countries established in 1958, initially known as the EEC, as the first step to the third form of integration, a common market, in which restrictions on trade and factor movements (capital and labour) are abolished. But from the beginning, the EUR6 aimed to go even further, at least to Balassa's fourth category: an economic union. In such a union there is some harmonization of economic policies to remove discrimination which exists if policies differ. Finally, total economic integration implies the establishment of supranational bodies so that monetary, fiscal and social policies are unified across the territory of the union. At the present time, the EU is in a transition phase, seeking to complete the third stage and moving towards stages four and five; it was the nature of this transition, and the characteristics of the final goal, which formed the subject of the 1991 Maastricht summit of EU leaders and has since been the cause of considerable uncertainty within Europe.

Common to all these forms of economic integration is the dismantling of at least some barriers to trade among the participating countries – tariffs, quotas and other non-tariff obstructions. The purpose is to facilitate trade and hence to raise incomes in the participating countries and encourage a higher rate of economic development than would otherwise occur. The primary effect of reducing or eliminating tariff and other barriers to trade is to lower the cost of transactions which take place across national frontiers. These lower costs may be reflected either in higher profit margins for the firms involved or lower prices to consumers, or some combination of these two responses. What actually happens depends on numerous factors, but especially on how competitive the relevant markets are. The general expectation is that most, if not all, of the cost reduction will be reflected in lower prices. Price changes will have an impact on the volume of trade, and the combination of price and volume changes will in turn affect the trade balances of the participating countries. Price changes also affect the income of consumers and hence their consumption patterns. All of these changes have implications for the scale of output, which has further consequences for prices, location of output and the patterns of trade. These are complex changes which are hard to analyse. In practice, much of the analytical work has focused on the short-run, or static, effects of economic union, mainly the trade effects of tariff reductions, on the assumption that cost reductions are passed on and result in price reductions. Important and necessary though these studies are, they are incomplete if the full dynamic consequences of economic integration are not taken into account.

TRADE DIVERSION AND TRADE CREATION: THEORY

Orthodox trade theory focuses on the idea of comparative advantage based on the relative abundance of the factors of production – land, labour and capital. If there are no barriers to trade it is possible for the economic system as a whole to obtain the maximum benefits of production specialization according to the comparative advantage of the respective trade partners. To the extent that trade barriers continue to exist, the opportunities for reaping these benefits are diminished. Trade barriers include transport costs, tariffs and a multitude of non-tariff impediments. Therefore, if in the initial situation there are tariff or non-tariff trade barriers, their reduction or removal will create conditions in which a reallocation of production will be possible and advantageous. Such a reallocation provides the basis for the *static* gains to be obtained from economic integration.

Figure 4.1 illustrates the trade effects expected from economic integration. In the initial situation, countries A and B do conduct trade with each other, but trade with third parties is twice as important. If the two countries eliminate mutual trade restrictions, but at the same time maintain the pre-existing trade regulations with third parties, two effects may be expected, shown schematically in the lower half of Figure 4.1. The trade of A and B with third parties has been halved, while that between them has been trebled. Some of the increase in trade between A and B is the direct substitution of sources of supply; over a range of products, A and B find that their partners are now cheaper sources than are countries outside the area of economic integration. This substitution effect is known as trade diversion. In the second place, the elimination of trade restrictions may open up production and trade possibilities which did not exist previously, thereby expanding economic opportunities. As a result, trade creation occurs (Viner 1950).

The static effects of economic integration amount to the short-term opportunities for reallocating production and trade geographically. In addition, one may expect dynamic effects to occur over the medium to long term, for at least two separate reasons. If the isolation of national markets is reduced or eliminated, manufacturers may be able to realize economies of scale by concentrating production in a limited number of plants which will supply both (all) countries. The second source of dynamic gains arises from the increase in competition which is implicit in the reduction of trade barriers. If firms find that competition is more intense they will have to innovate more energetically to survive. An increase in innovation should lead to higher total output and an increase in intra-industry trade. The dynamic effects arising from scale economies and a greater tempo of innovation cannot be achieved immediately, and should be continuing benefits over the long run.

The trade effects which are schematically represented in Figure 4.1 clearly

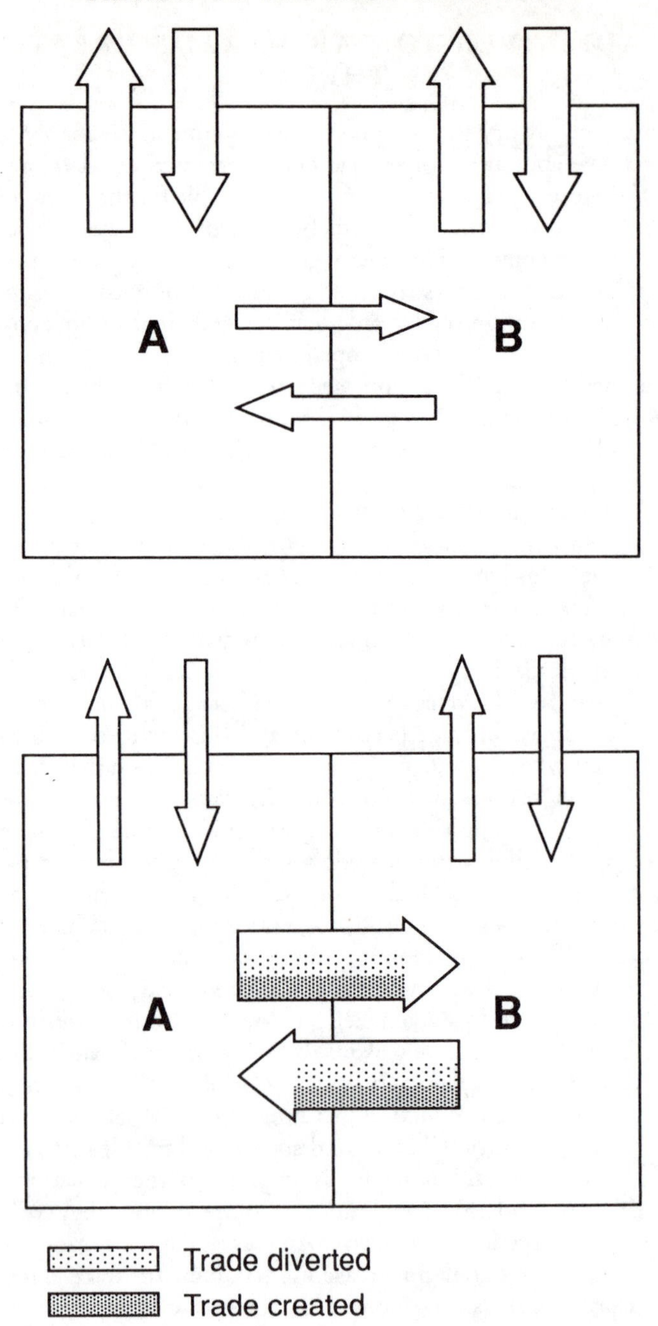

Figure 4.1 Trade patterns between countries A and B before the formation of a customs union (upper) and after (lower)

have geographical implications. First, the participating countries may be differentially affected by shifts in trade patterns and associated changes in production. Abstract theory provides no basis for predictions on this matter, the outcome depending on the circumstances of the participating countries. Second, integration will have intra-national effects, the direction of which can be readily deduced even though the magnitude depends upon the particular circumstances. Consider firms whose location choices are determined primarily by their access to component inputs and by access to their markets, rather than by the location of localized natural resources and/or labour supplies. For these firms, a change in trade pattern of the kind depicted in Figure 4.1 implies a change in the relative advantages of locations near the external frontiers and those along the frontier which separates A and B. Where it is access to markets which dominates location decisions, rather than access to raw materials, fuel and labour, locational advantages will shift in favour of those regions with good access to the whole territory of the integrated area, i.e., towards the frontier zone between the two participating countries depicted in Figure 4.1 (Giersch 1949–50).

In the EUR12 context, and taking account of the relative sizes of the respective countries, the region which has gained most in locational advantages on account of economic integration lies along a roughly north–south alignment, from the Netherlands to Marseilles–Milan. Within that general area, the greatest benefit will have been felt in the Benelux–Netherlands region, on account of the number of national frontiers located here and the fact that these three small countries lie between three of the larger economies (Britain, France and Germany). To the extent that the formation of the EU and its subsequent enlargement has resulted in trade diversion and trade creation, it is this part of the continent which should have benefited the most – from the increase in transit trade and from shifts in locational advantages for production. Furthermore, this area is, in general terms, that part of Europe which is often described as 'central' (see Chapter 5). Economic integration should, therefore, tend to foster growth and development in that part of Europe which is already favoured, to the detriment of peripheral areas (see Figures 1.1 and 1.2, and Table 1.1).

The theoretical arguments are clear and persuasive, that economic integration should raise the level of output and also accelerate the rate of growth in that output, for the group of participating countries. In addition, regions near national frontiers within the area encompassed by a union should gain differentially, leading to geographical shifts in relative prosperity. However, the magnitude of these effects will depend on the particular circumstances pertaining to a given union. Furthermore, measurement of the impact poses numerous technical and theoretical problems. Hence the fundamental question which must be addressed is whether the expectations derived from the theory of economic integration have been realized in practice? Much of the work which has been done in this field is concerned with tariff reductions and their impact on trade, and it is this body of evidence to which we will first turn. In addition, though, considerable effort has been expended to assess the probable impact of the SEM on the economies of the EUR12 countries, and this work will also be considered.

TARIFF AND NON-TARIFF TRADE IMPEDIMENTS

At the time the EU was formed by the EUR6 countries in 1958, the inter-war legacy was still evident in the high level of protection which characterized the global economy in general, including Europe. Even if we restrict ourselves to protection by means of tariffs, measurement of the level of shelter provided is difficult. Nevertheless, Table 4.1 gives some idea of the situation in Europe prior to the formation of the EU. This picture is confirmed by Nevin (1990: 104), who observed that by the time the transition period was over, tariffs between the member countries had fallen by 12 per cent (i.e., from an average of 12 per cent to zero). Clearly, tariff barriers in the 1950s were quite high, and economic integration could, therefore, be expected to have a substantial impact on the EUR6 countries and on their relationship with the wider world.

Successive rounds of GATT negotiation have served to reduce international tariff levels, with two effects. By 1991, the average external tariff on manufactured goods entering the EU (as enlarged to twelve countries) had fallen to 4 per cent, while the average for EFTA lay in the range 2–3 per cent (*Economist*, 26 October 1991: 105). The tariff effect of economic integration has clearly been much reduced over the post-war period, implying that the trade creating and trade diverting effects have declined. If we consider the EUR6 countries, for example, the worldwide reduction in tariffs which has

Table 4.1 Import duties in the 1950s

	c. *1952*[1] *Weighted average level of import duties*, ad valorem *(%)*	*1955*[2] *Tariff levels (%)*	
		Unweighted average	*Weighted average*
Germany	33.5	15.5	5.6
Italy	24.9	17.3	7.1
France	21.9	18.1	5.1
United Kingdom	12.4	–	–
Netherlands	11.0	9.5	5.5
Belgium–Luxembourg	10.2	9.5	4.3
Scandinavian countries	9.7	–	–

Sources: [1] Scitovsky 1958: 76
[2] Balassa 1962: 46

occurred over the last forty years will have diminished the impact of abolishing tariffs on intra-EUR6 trade. In addition, as the general level of tariffs has fallen, the impact to be expected from the accession of new members to the EU must also have become less. When Britain joined the EU in 1973, the United Kingdom tariff barrier was around 6 per cent, or about half that which existed among the EUR6 countries at the time the EU was formed in 1958 (Nevin 1990: 110).

Non-tariff barriers are a completely different story. There is no simple way in which to measure the trade-impeding effects of technical specifications, procurement policies, documentation requirements and the plethora of other matters which may serve to hinder imports into a country. As tariff barriers have been lowered, governments have been tempted to raise non-tariff impediments to trade, which actions can prove extremely difficult to identify and control. The GATT, and other bodies with a concern to increase the freedom of trade, have become increasingly concerned in recent years. However, it seems probable that, at least for manufactures, any increase in non-tariff protection has been too small to negate the impact of tariff reductions. Nevertheless, the significance of the issue is indicated by the much-heralded completion of the Single European Market. It was intended that, by the end of 1992, all non-tariff trade barriers should have been removed; the prospect was held out that this would serve to stimulate the EU economy quite dramatically. In addition, the 1993 Uruguay agreement of GATT includes provisions to control non-tariff impediments to trade, signalling the determination of the international community to maintain and increase genuine freedom to trade.

TRADE DIVERSION AND TRADE CREATION: EVIDENCE FOR THE ORIGINAL EU (EUR6)

Measurement of the impact of economic integration on trade flows, and hence on national growth rates, raises some formidable problems – both of theory and of empirical analysis. In the EU context, these problems have been well reviewed by Mayes (1978). Four main issues can be identified:

1. To measure the effects of economic integration, it is necessary to compare the actual outcome with what would have occurred in the absence of integration. Specification of the counterfactual situation is, to put it mildly, difficult.
2. It is desirable to distinguish between the short-term reallocation effects (static gains) and the dynamic long-run effects. Most studies concentrate on the short-run effects.
3. For the analysis to carry real conviction, a fully specified econometric model should be developed which takes account of supply and demand pressures, levels of economic activity, etc. This is an ambitious ideal which it is difficult to realize.

4 *Ex ante* assessments of the effects may differ from those using *ex post* data. In practice, most studies have been retrospective, using *ex post* data.

Most of the studies that are relevant for our purpose have tried to measure the aggregate effect of economic integration among the EUR6 countries when they formed the EU in 1958, with a transition period that was intended to be twelve years but which was somewhat curtailed in practice. The main feature of the 1958 union relevant in the present context was the abolition of tariffs on intra-EU trade. Scitovsky (1958) set the scene with his study of the impact that would follow from a union of the EUR6 countries, Britain and the Scandinavian states. He was thoroughly sceptical about the significance of the benefits to be obtained:

> Our numerical estimate of the gain from increased international specialization among members of a Western European union confirms the pessimism we voiced earlier concerning the scope for such specialization. It certainly seems that the textbook example of the benefit from increased trade and specialization is very unimportant indeed in the case of Western Europe.
>
> (Scitovsky 1958: 68)

Balassa (1962) was harshly critical of this work, believing that the gains from integration were greater than Scitovsky allowed, but the evidence which he adduced has limitations every bit as great as the evidence that he criticized.

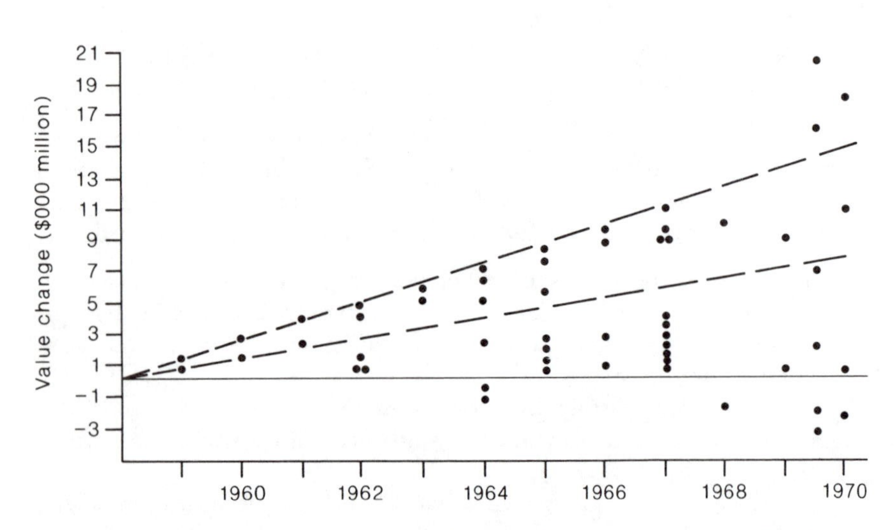

Figure 4.2 Predictions of trade creation plus trade diversion in the EU, collated for specific years from various sources
Source: Mayes 1978: 21

Some two decades later, Mayes (1978) reviewed the then extant literature on trade diversion and trade creation in the original union of six countries, prior to its first enlargement in 1973. Most of the twenty-one empirical studies on which he reports show that there was a positive impact on trade between the EUR6 countries, but the magnitude of the estimated effect varied considerably. Figure 4.2 shows the estimated trade impact in various years, collated from the studies reviewed by Mayes, measured in US$bn. The two pecked lines are the 'arbitrary bounds' identified by Mayes. He concludes:

> Several of the more extreme results lie outside these bounds, yet the range for 1970 is approximately $8–15,000 mn, so it is clear that we are a long way from achieving an agreed point estimate.
>
> (Mayes 1978: 20)

An alternative interpretation of Figure 4.2 is that, in 1970, the upper limit of the trade diversion and trade creation effects was about $20 billion, but that the impact could lie anywhere between that figure and a negative change of $3 billion. In the same year, the sum of imports plus exports for the EUR6 was about $180 billion. The maximum trade impact shown by these studies is therefore of the order of 11 per cent, but with the probability that it was a good deal less, possibly close to zero.

In a somewhat earlier paper, Balassa (1974) also reviewed a number of studies of the trade diversion and trade creation effects of the EU. The eight studies included in this review showed considerable variation in the magnitude of the impact, confirming Mayes's statement that an agreed point estimate is not available. However, Balassa concludes that the impact was positive and that trade creation effects substantially outweighed trade diversion; the main gains were identified in intra-industry trade creation, arising from the realization of scale economies and the rationalization of production. Overall, Balassa estimated the effects to be equivalent to 0.5 per cent of the 1970 GNP of the EU, at a time when GNP was growing at 5 per cent per annum (Balassa 1974: 127).

Other studies have made much stronger claims for the positive effect of the EU on the member countries:

> Once allowance is made for their inadequacies, the other studies tend to confirm our own results. Hence we conclude that intra-EEC trade in 1969 was something like 50 per cent greater than it would have been if the EEC had not been created.
>
> (Williamson and Bottrill 1971: 342)

A rather similar estimate has been provided by Nevin (1990), to the effect that by 1970 net trade creation from the formation of the EU amounted to between 28 and 34 per cent of the external trade of the EUR6 countries. He further suggests that there may be a dynamic benefit which amounts to an annual increment of 1 per cent of national income (Nevin 1990: 104–8).

Effects of this magnitude are hard to reconcile with the evidence reviewed in Chapter 2, showing a general increase in intra-trade among the EUR12 countries in the early post-war period, which appears to be part of a general increase in the relative importance of trade among the developed countries. The dynamic gains indicated by Nevin are also not consistent with the evidence of scale economies and their significance for intra-industry trade. Owen (1983) examined this issue, with special reference to France, Germany and Italy – the three giants among the EUR6 countries. His detailed data refer to the manufacture of cars, trucks and white goods (domestic appliances). For all three industries there was a substantial increase in the minimum efficient scale of production. Over the decade from 1970 to 1980, intra-trade expanded considerably faster than output. In total, Owen estimates that trade-related resource benefits amounted to between 3 and 6 per cent of EUR6 GDP by the year 1980 – estimates that are considerably lower than those offered by Nevin. Of equal interest, though, is Owen's conclusion about the slowness with which adjustments take place: 'The industry studies suggest that economic integration is a slow process, requiring 15–20 years to show measurable benefits and perhaps 40–50 years for completion' (Owen 1983: 158).

There is general agreement that the effect of the EU on trade and output in the member countries has been positive, but there are major discrepancies over the estimated impact of economic integration. These differences stem from the abiding problem of how to estimate the counterfactual situation, what would have happened in the absence of trade liberalization among the EU countries. However the issue is dressed up, the heart of an assessment is a comparison of what has actually happened, which is known with reasonable certainty, with some hypothetical, and essentially unknowable, alternative that would have been the actual state of affairs in the absence of economic integration (for example, Balassa 1989; Mendes 1987; Millington 1988). The evidence discussed in Chapter 2 suggests that there have been powerful forces pushing in the direction of greater trade among the industrialized countries, forces which would have operated even if the EU had not been formed. At this stage in the discussion, two points need to be emphasized. First, the main reason for the differing estimates of the effect of economic integration lies in the problem of specifying the counterfactual position. Second, given the strength of the trend to greater trade among the industrial countries – which is now much more clearly visible than it was 10–20 years ago – it is at least plausible to suggest that any bias in estimating the counterfactual situation would be towards understating the 'expected' volume of intra-trade, and therefore towards exaggerating the impact of economic integration.

TRADE DIVERSION AND TRADE CREATION: EVIDENCE FROM BRITAIN'S ACCESSION TO THE EU

Most of the EUR6 studies referred to in the last section span the period from 1958 to 1970. This period of twelve years includes the transition phase during which tariff barriers were dismantled, this transition lasting for ten and a half years. In contrast, most of the studies of Britain's entry into the EU in 1973 cover the period up to 1979, including the much shorter transition period allowed for than was agreed for the original union. It is a reasonable proposition, therefore, that the observable effects for Britain after six years should be much less than for EUR6 after twelve. In addition, the average tariff barrier for the EUR6 countries in 1958 was about 12 per cent, whereas the barrier which existed in 1973 between Britain and the EUR6 countries was about 6 per cent. The combination of shorter time period and lower level of tariffs should mean that the measurable impact of Britain's accession would be much less than in the case of the original EU. At the simplest and most naïve level, one might expect the measured impact for Britain to be only one-quarter of that for the original EU (half the time period multiplied by half the tariff level).

Mayes (1978: 22–3) reports three *ex ante* estimates for the static effects for Britain of the enlargement of the EU in 1973. The effects are calculated for the increase in the volume of manufactured imports into Britain by 1977/8 attributable to membership. These estimates range from an increase of $391 million to $2,559 million, at prices which varied from 1968 to 1972. In the year 1970, total manufactured imports amounted to $11,195 million. Therefore, by 1977/8, the static effect on imports of manufactures was of the order 3.5 to 23 per cent of the 1970 import volume. Among these three studies there clearly is considerable divergence of testimony, an absence of agreement that is similar to the situation already reported for the EUR6.

Two other studies, both *ex post* in nature, provide estimates of the impact on Britain of joining the EU. Winters (1987) examines the shortcomings of previous studies and then offers his own estimates as the best available. He estimated the share of Britain's trade with each of the EUR6 countries that could be attributed to her joining the EU in 1973. On this basis, he calculated the percentage by which trade in manufactures was larger than it otherwise would have been. The highest of these increases were imports from the Benelux countries (89 per cent) and exports to Germany (57 per cent); the lowest increases being imports from France (62 per cent) and exports to France (25 per cent) (Winters 1987: Table 6). These percentage increases imply an adverse impact on the balance of trade in manufactures, an implication which is supported by Fetherston, Moore and Rhodes (1979, 1980). They consider that the overall impact on trade in manufactures over the period 1973 to 1977 was negative to the extent of £1.1 billion annually at

1970 prices. They conclude that, as a result, national income was almost 12 per cent less than it would have been had Britain stayed outside the Union.

These estimates appear to be high in comparison with the estimates already mentioned for the EUR6 countries and given the *a priori* expectation that the measurable impact on Britain would be less than for EUR6 on account of the shorter period over which the assessments have been made and the lower initial tariff hurdle. Further doubt is created by the clear evidence presented in Chapter 2 that trade between Britain and the EUR6 countries has expanded rapidly and consistently over most of the post-war period. These doubts are reinforced by the findings of the most thorough study to date of the impact on Britain of joining the EU. Millington (1988) examines British exports to the EUR6 countries in 1970/1 and 1978/9. For this purpose, ninety-three manufacturing industries are identified, defined at the three-digit level of the SITC. To estimate the counterfactual situation had Britain not joined the EU, Millington adopts the following procedure. The main industrial-country markets outside Europe lie in Canada, Japan and the United States. The change in Britain's share of imports into these three markets (treated in aggregate) between 1970/1 and 1978/9 may be taken as indicative of the change in her import share of the EUR6 countries that would have taken place had she not joined the EU. Orthodox theory holds that Britain's export performance in trade with the EUR6 countries should have improved relative to performance in the markets of the other major industrial nations. However, Millington's results do not confirm these expectations:

> These results ... provide little support for the predictions of the traditional theory of customs union. The tariff variable is not significant in any of the estimated equations, and the measures of revealed comparative advantage are either not significant or have the wrong sign. The negative relationship between revealed comparative advantage and the dependent variable suggests that the UK has performed relatively badly within the EC(6) in those commodity markets in which it was strongest, relative to the world and the EC(6) in the pre-entry period.... The analysis suggests that the entry of the UK into the EC has not resulted in the relocation of productive capacity in the low cost centres of production, in line with the pattern of comparative advantage.
>
> (Millington 1988: 76–7, 113)

Millington then examines the evolution of intra-industry trade in manufactures between Britain, the EUR6 countries and the three countries mentioned in the previous paragraph – Canada, Japan and the United States. The hypothesis is that accession to the EU should have encouraged intra-industry trade with the EUR6 countries. Prior to Britain's accession, intra-industry trade accounted for about 76 per cent of trade in manufactures with

EUR6, and this proportion remained virtually constant thereafter. In contrast, there was some increase in the importance of intra-industry trade with the other three trade partners.

The counterfactual situation obtained by Millington is carefully specified and seems to be more plausible than those used in other studies. The small, almost negligible, impact of accession on Britain's trade flows is much more nearly consistent with the results obtained for the EUR6 countries than are the very large estimates obtained by some other scholars. Indeed, Millington's results are entirely consistent with the evidence contained in Figures 2.5, 2.6 and 2.8. The first two, it will be recalled, show the EUR6 countries' share in Britain's imports and exports respectively. The long-term rate of growth in these shares seems to have been little affected by the formation of the EU in 1958 without Britain, the formation of EFTA in 1960 as a rival economic bloc, or Britain's accession to the EU in 1973. If there were an impact on trade anywhere near that suggested by Mayes or especially by Winters, a distinct break in the trends would be expected. Similarly, Figure 2.8 shows quite clearly that the change in balance of trade in manufactures with the EUR6 countries has been very similar to the change that has occurred in the balance with the rest of the world. This conclusion is consistent with the evidence discussed by El-Agraa (1984), which suggests that the measurable impact of Britain's accession to the EU was small but more likely to have been positive rather than negative.

EUROPE 1992

From the time that the EU Commission published a White Paper in 1985 on the completion of the internal market until early in 1993, by which time it was intended that the programme should have been completed, there was considerable discussion of the probable (*ex ante*) impact of the SEM. An important component of that discussion was a major study which the Commission put in hand, the results of which were presented in a popular format by Cecchini (1988); the more scholarly version was published by Emerson *et al.*, also in 1988. Recall that by 1985 tariff barriers to trade among the EUR12 nations had been eliminated, so that the task being undertaken by the end of 1992 was to dismantle the non-tariff barriers to trade, to which end some 300 proposals had been put forward. Emerson and his team sought to evaluate the impact of moving to the SEM, and in particular to estimate the scale of the overall benefit to be obtained.

At the time this study was undertaken there was already clear evidence that Europe was slipping in the international competitive stakes, and one of the motives for completing the internal market was very precisely to improve the competitive position of Europe in the global economy:

> The potential gains from a full, competitive integration of the internal

> market are not trivial in macroeconomic terms. They could be about large enough to make the difference between a disappointing and very satisfactory economic performance for the community economy as a whole.
>
> (Emerson *et al.* 1988: 6)

That assessment was based on the estimate that the annual growth of output and consumption in the period up to 1992 could be raised by about 1 per cent for the EU as a whole, and that once these static gains had been realized 'longer-run dynamic effects could sustain a buoyant growth rate further into the 1990s' (Emerson *et al.* 1988: 5).

Without doubt, this study is the most thorough and comprehensive that has been undertaken to examine the effects of removing non-tariff barriers to trade. The authors use both macro-economic and micro-economic techniques to evaluate the impact of removing frontier formalities, harmonizing technical specifications and indirect taxes, opening up public procurement to international competition and the other measures provided for in the SEM. In addition, an attempt was made to assess the dynamic effects from realizing economies of scale and a more competitive environment for firms.

A clear distinction is made between an assessment of the 'narrow, technical, and short-term' benefits of the SEM and those which are to be expected to accrue from 'having a fully integrated, competitive and rationalized internal market' (Emerson *et al.* 1988: 7). The gains estimated under the latter heading are thought to be twice those available under the former. To achieve the full advantages of a fully integrated market, certain conditions would have to be met in both micro-economic and macro-economic policy, the net effect of which would be to convince firms that the SEM really is intended to operate and that policies will be rigorously implemented to ensure that freedom of passage across frontiers becomes a reality, that distorting subsidies will be abolished, etc. However, a more recent work, reviewing the above study and other recent publications, casts doubt on the robustness of the consensus that the SEM would have a substantial impact, even assuming that it were fully implemented: 'Comforting though the consensus is, it is built on relatively shaky foundations' (Winters and Venables 1991: 1; see also Mayes 1993; Winters 1992).

It is too soon to know whether the SEM has had the effect which was desired and intended. However, the anecdotal evidence is that the SEM, which was supposed to be complete by the end of 1992, is far from functioning as intended. Furthermore, the concept of the SEM implies that the EUR12 will have a common trade policy with the rest of the world, but there is relatively little sign that this will be achieved. In the context of mutual trade sanctions between the EU and the USA (initiated by the USA on the grounds of unfair competition), Germany and the USA have agreed to establish their own trade policy for public procurement in the field of telecommunications equipment. Meantime, the French persisted until the bitter end in trying to block an agreement under the Uruguay Round of GATT which they believed would be harmful to their agricultural interests. Finally, the economic prospects for the EUR12 countries in 1993 and 1994, although improving, are hardly buoyant, as the long and painful recession has eased but slowly. If the SEM really has yielded the promised benefits, these have been cruelly snatched away by other events of even greater moment, such as the unification of the two Germanies and the consequential pressure on investment resources and interest rates (*Economist*, 19 June 1993: 16). For the present, it is not clear whether the SEM really has yielded the promised benefits. At the very least, the experience to 1994 suggests that in practice these benefits may be rather small compared with other factors which affect the prosperity of nations. This conclusion is consistent with the conclusions reached in reviewing the earlier studies of the EUR6 countries and of Britain's accession to the EU in 1973.

CONCLUSION ON THE MACRO-EFFECTS OF ECONOMIC INTEGRATION

The preceding discussion has shown quite clearly that there is considerable disagreement over the magnitude of the impact attributable to economic integration, even though most observers conclude that there is a measurable and positive effect. Perhaps the main reason for this lack of agreement is the difficulty – both in theory and in practice – of specifying the counterfactual position. Reviewing the situation towards the end of the 1980s, Mayes came to the following conclusion:

> It is clear from the preceding discussion that quantification of the effects of integration is relatively rudimentary partly due to the lack of consideration of the many problems involved, but more importantly due to the inherent complexity of the effects. While changes in tariffs may exert straightforward static effects on trading patterns, these effects are observed in a dynamic context of evolving trade, payments and activity. That evolution must be explained to attempt to measure the effects of integration. The use of the difference between actual behaviour and some hypothetical 'anti-monde', which might have occurred without integration, based on the extrapolation of previous trends and concurrent behaviour in other countries and markets is bound to result in biases because all residual changes in behaviour are attributed to integration, not just those which can be directly explained by it. Furthermore such analyses often exclude the feedback of reactions to tariff and relative price changes on incomes and efficiency. Given the very small size of the static effects on trade and payments compared with a change in the rate of economic growth by half of 1 per

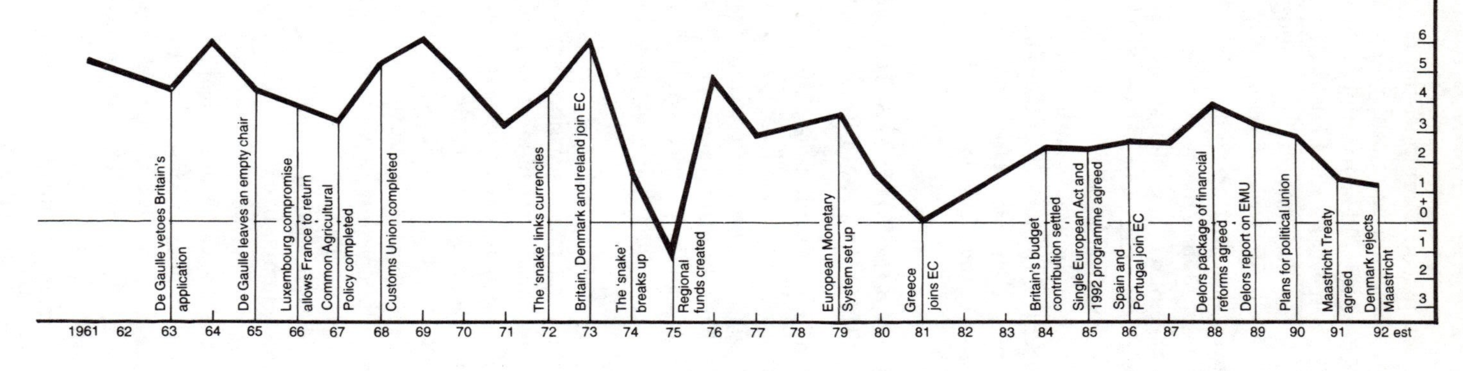

Figure 4.3 EUR12: annual percentage change in real GDP, 1961–92
Source: *Economist*, 19 December 1992: 19

> cent, it would be very easy to draw totally erroneous conclusions over the size or even the sign of the aggregate effect.
>
> As soon as the analysis is extended to include changes in economic growth, efficiency and the movement of factors rather than just trade in goods and services the problems of quantification become immense. There is no well developed theory of the interrelated movements of trade, payments, incomes and factors, let alone one which can be clearly applied and estimated to calculate the effects of integration.
>
> (Mayes 1988: 57)

Mayes went on to note that interest in quantification of the effects of economic integration had waned, though he might have noted the partial revival with the then pending introduction of the SEM. Nevertheless, it is abundantly clear that he is right to emphasize the difficulty of the task. It also seems highly probable that he is right in suggesting that the measurable impact of economic integration is small relative to other factors which determine economic performance. It certainly is the case that successive enlargements of the EU have not in fact been accompanied by an acceleration in the rate of growth by GDP, but rather the reverse (Figure 4.3). The long-run trend is clearly downward, and it was only in the case of accession by Greece in 1981 that there was subsequently an acceleration in growth rate. However, the Greek economy is small and both the scale and sustained nature of the upturn implies that other factors were at work. It seems quite clear that whatever the beneficial impact of economic union and of the SEM may be, the best that can be said for these effects is that in their absence the economic situation of the EU would have been either less good than, or worse than, it was in fact.

THE REGIONAL EFFECTS OF ECONOMIC INTEGRATION

If, as seems possible, the aggregate effects of economic integration are impossible to quantify and in any case may be quite small, we would expect that any regional impact would be even harder to detect. Indeed, one might suppose that the spatial consequences are so slight that they are not worth looking for even at the inter-national level, let alone the intra-national scale of analysis. In this context, it is instructive to note that whereas there has been a considerable literature on the macro-effects of economic integration, and on the impact on individual countries, there has been very little analytical work on the spatial implications more generally. This dearth of regional literature reflects, no doubt, the even greater difficulty that there is in dealing with the spatial impact of integration than with the aggregate effect. However, there are some studies relevant for our present enquiry; they do not provide any support for the expectation that economic integration will

Table 4.2 Estimated static impact on regional output of forming the EC and EFTA, four countries

	Unweighted percentage change in output			
	Regions showing increased output		*Regions showing decreased output*	
Regions in	*EC*	*EFTA*	*EC*	*EFTA*
Denmark	–	+2.248 (3)	–0.278 (3)	–
Germany	+0.572 (18)	–	–0.654 (17)	–0.158 (35)
Norway	–	+0.480 (6)	–0.398 (17)	–1.030 (11)
Sweden	–	+2.559 (17)	–0.248 (18)	–0.091 (1)

Note: Figures in parentheses indicate the number of regions
Source: Bröcker 1988: 276–7

lead to a concentration of activity in those regions which possess a large number of frontiers, and especially frontiers between major countries.

Bröcker (1988; see also Bröcker and Peschel 1988) has formulated the theory of economic integration, as put forward by Viner (1950), in a manner that is explicitly spatial. For this purpose, trade patterns are treated in terms of the gravity model (see Chapter 3). In this formulation, a reduction in tariffs can be equated with a reduction in the distance which separates trade partners. Therefore, a spatially differentiated reduction in the 'distance' variable should have an impact on both trade and output. Four countries are taken for study – Denmark, Germany, Norway and Sweden. These are divided into seventy-three regions; in addition, the rest of the world is divided into nineteen countries or groups of countries. For the analysis, thirty-six industries are identified, which together cover all of the manufacturing sector and part of the primary sector. It is assumed that 1970 represents an equilibrium year after the formation of the EU and EFTA. Given that assumption – which is more than somewhat heroic – the actual trade flows can be used to calibrate the doubly-constrained gravity model, from which estimates can be derived of what the trade flows (and hence output) would have been in 1970 in the absence of the EU, of EFTA and of both organizations. Tariff changes are treated as changes in distance, which have their impact through the beta exponent. The result is an estimate of the static effects of economic integration.

Table 4.2 summarizes Bröcker's results, as the estimated percentage change in regional output attributable to the separate formation of the EU and EFTA. On the face of it, these results suggest that the EU was generally favourable for the regions of Germany but had an adverse impact on the three Scandinavian countries, whereas the reverse is true for the formation of EFTA. But closer inspection shows that the impact was rather more mixed, especially in Germany and Norway. Furthermore, most of the 'spatial variation of integration effects ... results from varying sectoral compositions of regional output, but only to a minor extent from spatial variations within each sector (Bröcker 1988: 278). One of the fundamental reasons for the small impact attributable to integration lies in the low beta values obtained from the gravity model, implying a small impact on trade (and output) due to distance and hence changes in 'distance' occasioned by tariff reductions. Hence:

> Regarding the empirical results obtained so far, there is a lot of evidence to show that the adverse regional effects of integration feared at first have not emerged. Even though significant regional impacts of trade liberalisation have been observed to affect regions unequally, there is no sign of gains concentrating in the centres and losses in the periphery.
>
> (Bröcker and Peschel 1988: 149)

Neven (1990) approaches the distributional aspects of economic integration in an altogether different manner. His starting point is the estimate that completing the SEM in 1992 was expected to yield a static gain in basic consumption of 4 per cent and the problem he addresses is the distribution of this gain among the EUR12 countries. Neven does not approach this problem by attempting to develop a fully specified model; rather, he examines a number of key factors which will determine the geographical pattern of relative gain. The first step is to identify the relative importance of inter-industry and intra-industry trade in the overall trade of each country with its partners in EUR12. He finds that Greece and Portugal engage predominantly in inter-industry trade, whereas Belgium, Britain, France, Germany and the Netherlands engage mainly in intra-industry trade, mostly among themselves. Ireland, Italy and Spain occupy intermediate positions. Where inter-industry trade is strongly marked, traditional comparative advantage arguments are important – and especially the natural endowments and the role of unit labour costs. In general, it is the southern countries which stand to gain the most on this consideration. In contrast, the northern countries generally have a comparative advantage in industries which are highly intensive in their use of human capital, but they have to contend with high unit labour costs.

The second main issue examined by Neven is the potential gains to be realized from economies of scale. Two sets of evidence are marshalled. First, 'engineering' data allow one to assess the minimum economic scale of plant and the proportion of national output that is obtained from plants of a smaller size in selected industries. Second, comparisons can be made of the distribution of employment within countries by size of firm. The conclusion which is derived from this evidence is that the northern countries generally have little scope to realize further economies of scale, whereas the southern countries have a considerable potential gain to exploit. This conclusion is partially confirmed by Caballero and Lyons (1991). They examined Britain, Germany and France and found that in these countries there are few, if any, gains from internal scale economies to be obtained from the completion of the SEM.

On the assumption that the SEM will indeed yield significant benefits for EUR12, Neven concludes as follows:

> The main conclusion which emerges from this study is, therefore, that Northern European countries should expect relatively small benefits from the completion of the Internal Market. These countries seem to be already well integrated with one another. This is indicated by the fact that cost differences across countries do not appear to be significant and by the fact that firms tend to be fairly similar in terms of size. At the same time, the Northern European countries should receive a relatively small share of the benefits accruing from better exploitation of comparative advantage between the North and the South of Europe. In the North, the UK stands out to gain most, being relatively less integrated. The main beneficiaries of the 1992 programme are, however, likely to be the Southern European countries, both in terms of exploiting comparative advantage and in terms of exhausting scale economies.
>
> (Neven 1990: 46)

CONCLUSION

The theory of economic integration is persuasive: that dismantling trade barriers should have a substantial impact on trade and output. However, the problems of finding suitable empirical tests to ascertain whether theoretical expectations are in fact fulfilled, and measuring the magnitude of those effects, are difficult to solve. As a result, there is no agreement on the magnitude of the point estimates. Although some rather extravagant claims have been made for the size of integration effects, the evidence from the formal assessments, set in the context of the evidence reviewed in Chapter 2, suggests that in fact the measurable consequences of integration are rather small, and very small in relation to other influences on trade and output. Even if GDP grows at only 1 or 2 per cent annually, it takes only a very few years to yield an increment in GDP equivalent to the estimated static benefits of greater economic integration. If that is indeed the case, then the spatial significance of economic integration is likely to be even harder to detect than is the aggregate impact. In which case, the argument that further economic integration will accentuate centralization tendencies has a very limited value.

This scepticism leads one to concur with El-Agraa (1900a: 79) that the real significance of economic integration may lie in the political, not the economic, arena. In the case of the EU, the real driving force for its formation was the passionate desire to ensure that never again would the countries of Western Europe bleed each other in armed conflict. This desire was strongly present in Europe itself (El-Agraa 1990b: 3, 12–13) and also in the United States. America actively assisted in the formation of the EU, helping to negotiate the breaches of GATT rules that were necessary for the EU to be formed, in the belief that the primary benefit of the EU lay in the political field (Bhagwati 1991: 69–70). An important aspect of this political dimension was the expectation that if the economies of the European nations were truly interdependent in the economic sphere, warfare between them would be impossible. Now it happens that until relatively recently the EU enjoyed high rates of economic growth. While it seems plausible to argue that this has been the beneficial result of the formation of the EU, the evidence reviewed in this chapter, and the divisions which have emerged in the recent years of recession, suggests an alternative interpretation. It may well have been the case that a rapid tempo of growth facilitated the economic and political adjustments entailed in the process of economic integration. After all, if everyone is getting richer it does not matter too much if the rate of improvement is unequally shared. When growth is slow or negative, any real or perceived redistribution effects take on a much greater salience.

The theory of economic integration leads one to expect that there will be clear and measurable benefits for the participating countries, and also that these benefits will be geographically concentrated in the 'central' areas within the union where there are many political frontiers. In practice, the expected benefits are hard to quantify and appear to be smaller than many would wish us to believe. If that conclusion is accepted, it follows that peripheral regions – or countries – have little to fear from economic integration solely on account of their relative location.

5

CUMULATIVE CAUSATION OR NEO-CLASSICAL PROCESSES?

> Unto every one that hath shall be given, and he shall have abundance: but from him that hath not shall be taken away even that which he hath.
>
> (St Matthew)

In the previous chapter, we have examined the available evidence regarding the impact of economic integration, finding that measurement is difficult but that, so far as may be judged, the impact has probably been rather small. However, that assessment is based on studies which, for the most part, consider only the static effects of integration. It is possible that the dynamic effects may be at least as important, if not more so, but are impossible to identify because they are manifest over a long period of time and are, therefore, inextricably linked with other dynamic processes. Therefore in this chapter we will examine two more general bodies of theory and some empirical evidence which may help us to discriminate between them. In shorthand terms, we may label these theories of regional growth as the cumulative causation and the neo-classical models. In essence, the neo-classical view argues that spatial movements of the factors of production, and trade in commodities and services, serve to equalize incomes between regions. This expectation is contested by the cumulative growth theorists, whose argument is encapsulated in the quotation from St Matthew. If the neo-classical view is correct, then peripheral regions have nothing to fear from the fact of their location, whereas if cumulative growth processes dominate, they are in danger of losing out to the central areas which enjoy the benefits of cumulative growth. In a recent general review of these issues, Chisholm (1990b) concluded that elements of both processes can be recognized in the advanced economies. However, it is appropriate to consider the matter in more detail and with specific reference to Western Europe.

NEO-CLASSICAL REGIONAL GROWTH THEORY

Neo-classical thinking has its roots in microeconomic analysis, i.e., the analysis of the behaviour of individuals and of firms. To develop a coherent theory of economic growth, whether for the national economy or for regions, requires some heroic assumptions, including the idea that an economic system will move towards an equilibrium position in which the factors of production, and especially labour, are fully employed. In the long run, the rate of economic growth is assumed to be determined by two exogenous factors: the rate of population growth and the rate of technical progress. For the long-run view of regional growth, neo-classical thinking assumes away many of the interesting problems. On the other hand, it does offer some important insights into the process of adjustment from an existing disequilibrium situation towards the postulated long-run equilibrium-growth path.

If we assume that individuals and firms behave in a rational manner to maximize, respectively, their incomes (or utility) and their profits, and if we further assume that these micro-economic actors are blessed with full information about the options available to them and that transitions from one state to another are costless, then labour and capital will be allocated in a manner that will equalize returns among sectors and regions. Given these assumptions, the following processes may be visualized as applying to any system of regions in which, in the initial state, the returns to labour and capital are not equal. In one or more regions there may be a large number of people unemployed or underemployed, and as a consequence wage rates will be low. For most businesses, wages and salaries are the biggest single item of cost. A firm which sells over a substantial geographical area will therefore find its production costs lower in the low-wage region(s) than elsewhere, with the result that it will enjoy higher profits. These higher profits will encourage firms to invest in the low-wage region(s), encouraging a net inflow of capital into the low-wage economy. Meantime, workers are attracted to regions where there is less unemployment and/or higher wages, thereby generating a net migration stream towards the more prosperous regions.

Neo-classical thinking postulates that the movement of the factors of production in *opposite* directions will serve to equalize the profits of firms, bring unemployment in all regions down to the level which is accepted as 'full employment', and equalize wage rates. This theory of regional growth is, therefore, a theory of the transition from an initial disequilibrium to a long-term equilibrium, in which growth would be determined by population change and technological advances, both treated as exogenous processes (Chisholm 1990b).

Embodied in this theory are two further assumptions. First, even though a system of regions is being considered, it is assumed that there are zero costs associated with the spatial mobility of capital and labour, an assumption which is consistent with the intellectual origins of the theory in the concept of perfectly competitive markets but is not consistent with the manifest fact that transactions over distance do involve some costs, even if these are comparatively small (see Chapters 3 and 6). In principle, though, the fact that space does introduce an element of cost implies that full equalization of factor returns will not be achieved. The second assumption which is relevant in the present context is that there are no economies of scale. Neo-classical thought, therefore, assumes away the possibility that firms obtain scale economies in their own operations (internal scale economies) and may also benefit from the close juxtaposition of firms in space which yields external economies of scale, or agglomeration economies as they are often known. The omission of distance costs and scale economies is often seen as a fatal inadequacy of the neo-classical formulation of regional growth theory. In partial recognition of these difficulties, neo-classical thinking emphasizes the long-run nature of adjustment processes, which may take place over periods measured in decades (see, for example, Borts and Stein 1964).

CUMULATIVE GROWTH PROCESSES

Cumulative causation theorizing denies the tendency for regional systems to move towards an equilibrium at full employment, postulating instead a continuing process of growth in central areas which occurs to the detriment of peripheral regions. It is this body of thought that provides the main intellectual underpinning for the fear, noted in Chapter 1, that the geographical implications of close integration among the EUR12 economies will be adverse for Britain as a whole, and for the north and west in particular.

The basis for cumulative growth processes was set out independently by Hirschman (1958) and Myrdal (1957). Fundamental to this theory is the idea of scale economies, and especially external economies of scale (Clark 1967; Young 1928; Youngson 1967). If a region acquires some major productive investment and associated infrastructure, the scene is set for cumulative growth. The existence of the infrastructure – transport, electricity, water supplies, etc. – creates favourable circumstances for other firms to locate in the same area; they can enjoy external economies of scale in the provision of basic services. As the number of firms located in the area becomes larger, it will also be possible for specialist firms to be established that will provide material or non-material inputs to other firms, thereby accentuating the general advantages enjoyed by the area, and these advantages will be reinforced by the development of education, cultural and other facilities.

The immediate effect of obtaining inputs at lower cost is that firms in the region can obtain a return on capital that is higher than can be earned elsewhere. As a consequence, existing firms will be encouraged to expand their activities, and inward investment will bring additional firms to the region. Consequently, there will be a net inward flow of capital. With rapid capital accumulation, the average age of the capital stock will be lower than in other regions and, it may be assumed, this younger capital will be more efficient than the equivalent capital elsewhere. According to Verdoorn's proposition, therefore, the initial impulse will be accentuated and maintained by relatively high capital productivity.

The net inward flow of capital implies pressure on the locally available supply of labour, which in turn implies an upward movement of wages. As wages rise above those available in less fortunate regions, there should be a net flow of workers from the low-wage regions to the high-wage region(s). So long as the rise in wage costs is not sufficient to absorb all of the extra profitability which arises from the economies associated with the spatial concentration of activities, the situation can be maintained in which the favoured area offers higher returns to both capital and labour than are available elsewhere. Under these circumstances, capital and labour simultaneously migrate in the *same* direction and there is no tendency to move towards equilibrium, but rather towards the maintenance or even accentuation of disequilibrium.

Once a process of cumulative growth has become established, accelerator and multiplier effects will maintain and enhance the pattern of regionally differentiated growth rates. The accelerator effect arises from the migration of workers. When they move, their income is transferred from the region of out-migration to the region of migration gain. This effect is, therefore, directly proportional to the amount of income transferred. Multipliers accentuate this effect, in that the income of migrant workers, when it is spent, creates further income and employment. As a result, the total adverse income and employment impact in the region of outmigration is greater than the direct loss, while the increment in the region of in-migration is larger than the immediate gain.

The cumulative causation model of economic development was formulated in the Keynesian framework of disequilibrium economic processes (Chisholm 1990b) and was given considerable impetus by the work of Kaldor (see his 1972 paper in particular). However, Hirschman, Myrdal and others have recognized the possibility that the disequilibrium of cumulative growth may not continue indefinitely. Growth in the core should mean a rising demand for foodstuffs and other materials supplied from the periphery, and this demand may be sufficient to trigger rapid growth in at least some places outside the core, although this mechanism does not look very plausible in the European context. In addition, scale economies in the core itself may be exhausted, most notably through traffic congestion, high land prices and degradation of the environment. As diseconomies of scale set in, the attractions of the core region will decline and it may lose the momentum of development. However, most workers using the cumulative growth model appear to assume that if economies of scale at the regional level are in fact U-shaped, then in practice the regions of interest lie on the left-hand limb, so that further growth will be accompanied by a further reduction in costs and therefore accentuation of the cumulative growth process, with the implication that scale economies are not exhausted and that cumulative growth will continue.

Implicit in the proposition that external economies of scale are important in regional growth processes is the assumption that distance does impose costs on transactions of all kinds, an assumption which contradicts neo-classical thinking but which clearly accords with the observed reality that there is at least some friction of distance. On the other hand, the theory is unspecific as to the dimensions of the geographical area over which cumulative causation processes may operate; do they work at the scale of an individual city, over an entire region of a size such as South East England, or over an even bigger area? Without an assessment of the magnitude of the externality effects, and associated internal economies of scale, related to the actual cost of transactions over space, there is no means by which relevant judgements can be made. In the present context, though, it is the regional and international scales which are of particular interest.

A common way of approaching the practical problem of defining the regions which should gain from agglomeration economies is to prepare maps of generalized accessibility, on the assumption that the more accessible that a region is the greater the economies that can be achieved and hence the larger the potential for cumulative growth to occur. For this purpose, we may turn to the concept of economic potential. Economic potential is a concept in the same intellectual tradition as the gravity model (see Chapter 3). To calculate economic potential is straightforward in principle but requires quite a lot of computational power. Some measure must be selected for the 'mass' of a region – the value of GDP, total population or volume of retail sales, for example. Given data on the economic mass for every region in the geographical area of concern, and a matrix of the distances which separate them, the economic potential for any one region is obtained as the sum of the mass of each other region divided by the distance to it. The larger the resulting aggregate, the closer is a region to all the other regions considered – it has high potential accessibility. Conversely, a small aggregate indicates remoteness from other centres of population or economic activity. In effect, the areas with high economic potential may be regarded as those which benefit from economies of scale, and which will therefore benefit from the cumulative growth processes that have already been described.

The idea of generalized accessibility has been familiar since at least 1938 (Chisholm 1990b: 93), but was not popularized until the 1960s (Clark 1966; Clark *et al.* 1969). The first map of economic potential for Western Europe was published in the late 1960s (Clark *et al.* 1969), at a time when enlargement of the EU was in prospect. Clark's work has been updated by Keeble, most recently for EUR12 in 1988 (Keeble *et al.* 1982, 1986, 1988); Figure 1.2 reproduces these findings. This map shows the generalized pattern of economic potential, using GDP data (measured in ECUs) for 166 regions in EUR12 and also taking account of the adjacent countries. This map shows a very clear pattern of high accessibility in the area focusing on Paris, London, Hamburg and Stuttgart, i.e., an area centred on Belgium, the Netherlands and Luxembourg which, as we saw in Chapter 4, is also an area that should have benefited relatively from the trade diversion and trade creation effects of economic integration. Within Britain, although London has an economic potential equal to that of the other highly accessible centres on the mainland, northern Scotland has values as low as Portugal, southern Spain, Sardinia and other peripheral regions of EUR12. In general, the regions of Britain which are considered remote or peripheral in the British context – the South West, Wales, northern England, Scotland and Northern Ireland – are also peripheral in the EUR12 context.

To make the concept of economic potential operational, it is usual to define the geographical area of particular interest, such as Britain or the whole territory of the EUR12 countries. Economic potential then measures the generalized accessibility to this space. However, most, if not all, regions will have transactions with countries that lie outside the defined area, and the fact that accessibility for these external transactions varies between regions will be ignored. In the second place, use of economic potential measures implies that all firms trade across the whole of the defined economic space, which manifestly is not true. Alternatively, one must argue that though it is only a limited number of firms which need access to the whole space, the location of these firms is critical because this will determine the location of other firms supplying inputs to the production process (Keeble *et al.* 1986: 19–20). But this in turn depends on the significance of transfer costs relative to other location factors and hence the nature of regional multipliers.

Keeble and his co-workers have expressed some doubts about the significance of economic potential as a predictor of regional growth (Keeble *et al.* 1982: 430), and these doubts have been shared by others (e.g. Bröcker 1988; Bröcker and Peschel 1988; Chisholm 1964, 1985a, 1990b; Vickerman and Flowerdew 1990). Nevertheless, as we have seen in Chapter 1, there is a substantial body of literature, amounting to a conventional wisdom, that relative accessibility is in practice an important factor influencing the investment decisions of firms, and that, as a consequence, a continuing process of centralization is inevitable.

EVALUATION OF THE COMPETING THEORIES OF REGIONAL GROWTH

It is now time to ask whether the available empirical evidence throws useful light on whether the one or the other theory of regional growth is the more appropriate for analysing the experience of the EUR12 countries. Data are available for changes over time in the geographical pattern of GDP per caput, of unemployment rates and of factor prices, as well as information on migration flows. These four topics provide the focus for the ensuing examination, which is partly based on the fourth periodic report on the regions issued by the Commission of the European Communities (1991), the work of Keeble and his colleagues (Keeble 1989; Keeble *et al.* 1981, 1986, 1988), and the Commission of the European Communities (1993).

Gross domestic product

Figure 5.1 shows very clearly that the five main regions of relative poverty in the EU lie on the southern periphery of the Community – Greece, Ireland, the Mezzogiorno of Italy, Portugal and Spain (see also Table 1.1). However, to postulate a direct causal link between GDP per caput and relative location within Europe would be overly simple for at least two reasons. First, the low incomes of the five regions have been a phenomenon of the European scene for a very long time – several centuries at least – and for reasons which include the exploitative colonial regime that Britain maintained in Ireland and the relative decline of Portugal and Spain after the Age of Discovery. Second, and closely related, the present-day pattern is also an artefact of the

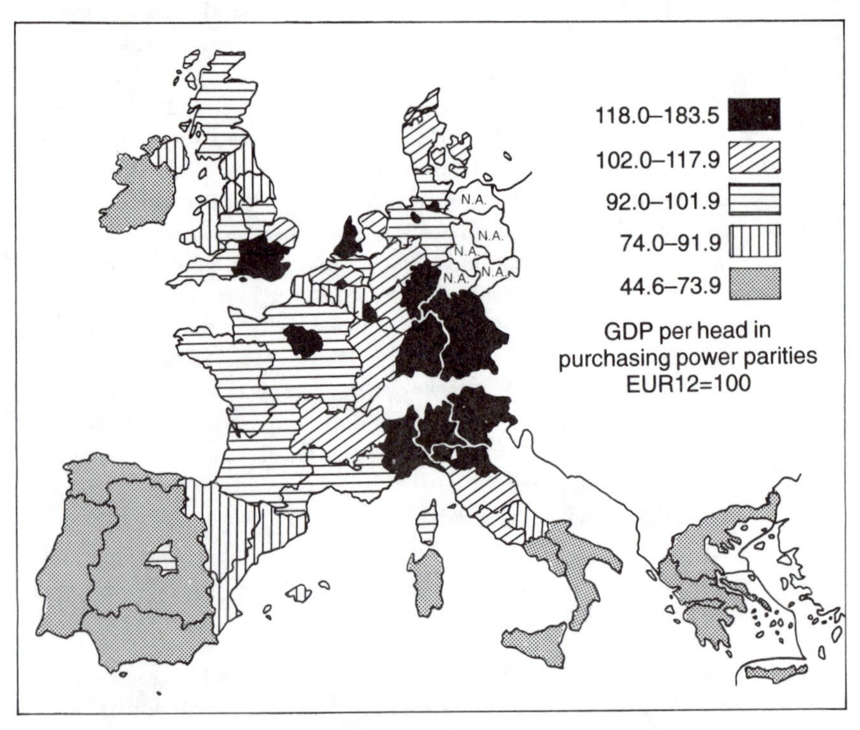

Figure 5.1 Income per head in the European Union, 1990
Source: D.E. Keeble, Department of Geography, University of Cambridge

expansion of the EU in the 1980s to include Greece, Portugal and Spain. Once Austria and the three Scandinavian countries complete the accession process in 1995, important countries whose per caput income is high will be included, but they will be peripheral within the Union.

The main interest is whether the poor regions, and especially those which occupy peripheral locations, have succeeded in narrowing the gap with the richer core regions. In reviewing the available studies, it quickly becomes apparent that whether income convergence or divergence is identified depends very much on the period of time over which an analysis extends, and also whether the analysis is conducted using national data or data for the constituent regions. One of the first studies found some evidence to suggest that peripheral regions were slipping back relative to the core regions of the EU, in the period 1973–7, but that from 1977 to 1983 the growth in GDP per caput was faster in the periphery than in the core (Keeble *et al.* 1981, 1988). Data compiled by the Commission of the European Communities (1991: 21) reveal a rather more complex pattern during the 1980s. Using a population-weighted coefficient of variation (standard deviation weighted by population divided by mean), disparities in GDP per caput using sub-national regions widened somewhat between 1980 and 1986, and were roughly constant thereafter to 1990. In contrast, disparities measured on the basis of whole nations, having also widened until 1986, then fell back to end the decade below the level in 1980. This implies that from the mid-1980s, two separate processes were at work: at the international level disparities became smaller on account of relatively favourable growth rates in the poorer countries; on the other hand, intra-national regional disparities increased, probably on account of the geographically differential impact of industrial restructuring (Dunford 1993).

In the present context, the most relevant scale of consideration is at the national level. Analysing EUR12 data from 1960 to 1992, Dunford (1994) shows that there was a remarkably strong and persistent closure in disparities at the national level between 1960 and 1974, that disparities then widened to about 1984 and have since returned to near the 1974 level, thereby confirming the findings of the Commission for the European Communities for the last decade or so. Broadly speaking, therefore, the post-war period falls into two parts. Pre-1974, disparities in GDP per caput fell sharply and continuously. Since 1974, disparities have remained much less than they were prior to 1965, but have not displayed an identifiable long-term trend. If cumulative growth processes had dominated the evolution of the EUR12 economies then we would have expected to see a persistent widening of disparities in GDP per person. The fact that there is no evidence that this is the long-term trend, and some evidence showing substantial convergence, implies that centralization processes have not, at the national level, had the impact predicted by cumulative growth theory.

Table 5.1 provides information for the individual EUR12 countries from

Table 5.1 EUR12 countries, GDP at market prices per head of population (EUR12 = 100)

	1960	*1970*	*1980*	*1990*	*1991*	*1992*	*1993*
Denmark	123.5	140.1	131.4	136.7	132.8	133.3	138.9
Luxembourg	158.2	141.7	126.7	127.9	127.1	129.6	137.4
Germany	123.5	132.8	133.6	128.6	129.2	131.7	135.2
France	126.6	122.9	125.3	114.2	110.5	112.6	116.9
Belgium	115.6	114.2	121.8	104.9	103.7	105.6	110.4
Netherlands	99.0	114.9	124.6	103.1	101.5	102.1	106.0
Italy	75.0	87.5	81.5	103.1	104.7	102.5	96.2
United Kingdom	131.3	97.2	96.8	92.4	92.2	86.9	82.2
Spain	36.6	49.0	58.2	68.5	71.0	70.9	69.5
Ireland	59.2	57.6	57.4	66.7	64.8	65.9	68.9
Portugal	28.0	32.0	27.4	33.1	36.7	43.5	46.3
Greece	40.0	49.5	42.1	35.7	36.1	36.8	37.5

Note: Countries arranged in descending order on the 1993 data
Source: *European Economy* 54, 1993: 189

1960 onwards. Although the relativities in GDP per caput have varied in the period shown, there is nevertheless considerable stability in the system. The four countries at the top of the table on the 1993 data were also among the five richest countries in 1960, but their relative position has changed considerably: Denmark and Germany have enhanced their relative positions, while France and Luxembourg have slipped significantly. In contrast, the United Kingdom, the second wealthiest country in 1960 now lies in eighth position. Belgium and the Netherlands have maintained their position near the EUR12 average, albeit with fluctuations. At the bottom of the table, Italy, Spain, Portugal and Ireland have all improved their position quite substantially relative to the mean for the EUR12 countries, with only Greece struggling to maintain even the status it had in 1960. The unweighted mean deviation from the EUR12 average was 36.7 per cent in 1960 and this had closed to 28.7 per cent in 1993. The clear impression given by Table 5.1 is that there has been some convergence in GDP per caput over the three-decade period, a finding which is consistent with the other evidence we have already reviewed. The really striking feature, though, has been the relative improvement of three out of the four poorest countries, all of which are peripheral in EUR12.

This evidence certainly suggests that cumulative growth processes at the international level are not dominating the evolution of Europe's space economy. It further suggests that the neo-classical model may in fact be operating. However, both of these conclusions must be treated as preliminary, conditional on what other evidence shows.

Migration

The political upheavals in Eastern Europe which began in the late 1980s and have continued into the 1990s have been associated with substantial population movements which it is reasonable to treat as 'abnormal'. Table 5.2 presents data for the year 1988, which can be taken as recording the position at the end of a long period of 'normal' migration flows. This table shows that within the EUR12 countries, but excluding Italy, there were 12 million people whose country of residence and nationality differed, and it also shows that these foreigners were unequally distributed among the EUR12 countries. The low-income countries have comparatively few foreigners, which is consistent with the prolonged process of net emigration. In contrast, the richer countries have clearly attracted considerable numbers of non-nationals. In aggregate, these migration flows are consistent with the predictions of both neo-classical and cumulative causation theories of regional growth.

However, it is striking that the number of foreigners drawn from outside the EUR12 group of countries is nearly double the number who have moved within the EU. This implies that, notwithstanding liberalization of labour laws, there are still considerable obstacles to labour mobility within the Community, an implication which is confirmed by the low migration rates experienced in the EU compared with elsewhere. Annual net migration

Table 5.2 EUR12 1988: foreign nationals as percentage of total population

	Source of foreign nationals		
	EU countries	*Non-EU countries*	*Total*
Italy	n/a	n/a	n/a
Spain	0.2	0.1	0.3
Portugal	0.1	0.5	0.6
Greece	0.1	0.6	0.7
Denmark	0.4	1.3	1.7
Ireland	1.9	0.5	2.4
Netherlands	1.2	2.6	3.8
United Kingdom	1.4	2.6	4.0
France	2.6	4.4	7.0
Germany	2.2	5.0	7.2
Belgium	4.9	3.5	8.4
Luxembourg	24.1	2.5	26.6
EUR12	1.7	3.0	4.7
Foreign population, million	4.4	7.8	12.2

Source: Commission of the European Communities 1991: 89

between the sixty-four statistical regions of the EU averaged 0.4 per cent of the resident population over the period 1970–9, declining to 0.2 per cent during the half decade 1980–5. These rates are low by comparison with the United States; annual net migration between the fifty states (plus Washington, DC) averaged 0.8 per cent and 0.7 per cent for the two periods respectively (Commission of the European Communities 1990: 151).

Flanagan (1993) has examined the available migration data over the period from 1958 to 1980 for the EUR6 countries, distinguishing intra-EU migration and immigration flows into the EU. Intra-EUR6 migration has fluctuated but shows a distinct decline in absolute numbers, from about 200,000 p.a. in the 1960s to about half that level from the mid-1970s. There is no evidence to suggest that liberalization of labour laws has had any impact on the volume of intra-EUR6 migration. Immigration into the EUR6 from outside the EU shows a very different pattern. In the early 1960s, internal and external immigration flows were roughly equal and growing at about the same rate. However, from 1961 onwards, immigration from outside the EU continued to rise strongly until 1964, declined from then until 1967 and after that year rose to a peak in 1970. In that year, external immigration was nearly four times as important as internal migration but then quickly fell to the level of the internal flows. Flanagan's evidence shows quite clearly that a period of widening wage differentials in the EUR6 countries was associated with a sharp increase in migration from outside the EU, but a decline in intra-EU mobility. This seems fairly conclusively to confirm that the loosening of labour laws within the EU has not had much impact on migration flows.

It seems quite clear that cultural and linguistic barriers remain important, serving to limit the scale of labour mobility within EUR12 (Commission of the European Communities 1991: 29; Gordon and Thirlwall 1989). This places limits on the scope for cumulative growth processes to operate at the international scale within EUR12. At the same time, though, the operation of the neo-classical model is also affected, with implications for levels of unemployment and wage rates. Even though migration has not been on a scale sufficient to equalize unemployment rates and wage levels between the regions of EUR12, there is some evidence that regions of migration loss have gained in economic terms, even if their gain has not fully matched the gain experienced by the regions of net immigration (Kindleberger 1967; Macmillan 1982). Formal assessment of the impact of migration is a difficult enterprise, and Macmillan (1982: 267) is suitably cautious when he concludes: 'On balance it is probable that the labour importing countries have benefited more from labour migration than have the labour exporting countries.'

Unemployment

In the thirty years since 1964, unemployment has risen quite considerably in all the EUR12 countries (Table 5.3). Since the end of the 1970s, every country

Table 5.3 EUR12 countries, percentage unemployment 1964–93

	Average for period					
	1964–70	*1971–80*	*1981–90*	*1991*	*1992*	*1993*
Luxembourg	0	0.6	2.5	1.6	1.9	2.0
Portugal	2.5	5.1	7.1	4.1	4.8	5.4
Germany	0.7	2.2	6.0	4.2	4.5	6.0
Netherlands	1.0	4.4	10.2	7.0	6.7	7.6
Greece	5.0	2.2	7.1	7.7	7.7	8.5
Belgium	2.0	4.6	10.7	7.5	8.2	9.3
Denmark	1.0	3.7	7.6	8.9	9.5	9.5
Italy	5.0	6.1	9.5	10.2	10.2	10.6
France	2.0	4.1	9.2	9.5	10.1	10.8
United Kingdom	1.7	3.8	9.7	9.1	10.8	12.3
Ireland	5.5	7.7	15.7	16.2	17.8	19.2
Spain	2.7	5.4	18.5	16.3	18.0	19.5
EUR12	2.3	4.1	9.6	8.8	9.5	10.6

Note: Countries arranged in ascending order on the 1993 data
Source: *European Economy* 54, 1993: 184

other than Luxembourg has suffered a level of unemployment well above any reasonable definition of 'full-employment', so that it is not very meaningful to characterize Europe as having some countries of labour shortage and others where labour is in surplus; it has been a matter of the degree of surplus. Luxembourg is a very special case, in that somewhat over one-quarter of residents are non-nationals, many of whom are guest workers who may be sent home if no work is available (Table 5.2). If Luxembourg is set aside, the pattern shown in Table 5.3 remains rather complex. Some of the traditional labour surplus countries, most notably Ireland and Spain, have for long experienced unemployment levels well above the EUR12 average. In sharp contrast, both Greece and Portugal have equally persistently experienced unemployment rates very near or below the average. These data show very clearly that there is no simple relationship between unemployment rates and the location of a country on the periphery or near the core of the EU.

Table 5.3 also shows that national unemployment differentials have closed over the thirty years shown. Ireland's unemployment rate was nearly eight times that of Germany in the period 1964–70, whereas the range of highest and lowest rates (Luxembourg excluded) had fallen to less than half that in the following decade and has stayed in the range of three or four to one ever since. At the national level, there has been a very substantial closure in the unemployment rates in the period shown, but most of that reduction in disparities occurred in the early part of the period. The Commission of the European Communities (1991) has calculated regional disparities, on the basis of sub-national regional units for the period 1970 to 1990, using the coefficient of variation. The results show a stable pattern at about 2 per cent until 1978 and then a sharp rise to about 5 per cent in 1987, with but a modest fall since then. This substantial widening of regional disparities in recent years must be attributed to the widening of intra-national unemployment differentials, reflecting the impact within nations of the difficulties associated with recession and the high levels of unemployment which have been the general experience.

Migration may have had something to do with the closure of employment differentials between the EUR12 countries but, as we have already seen, migration movements within Europe are in fact rather small. It seems altogether more likely that the limited scale of migration is due in the first place to the linguistic and cultural barriers which exist and in the second place to the rather high levels of unemployment throughout EUR12, which substantially reduces the incentive to move. It does not appear likely that migration flows have had a big impact on national unemployment differentials; it is more likely that the reduction in unemployment differentials has been occasioned by the relative tempo of economic growth discussed in a previous section (see pp. 82–4).

Finally, the recession years of the early 1990s had a particularly severe impact on many service industries, at least in Britain. As a result, southern Britain has experienced much higher levels of unemployment than ever before in the post-war period. In 1993, London's unemployment rate was above the national average, and when the map of Assisted Areas was redrawn in July, parts of the capital were designated for the first time, as well as parts of East Anglia, Essex, Kent and Sussex plus the Isle of Wight. In other words, significant parts of the relatively accessible South East were afflicted with the unemployment symptoms more commonly associated with the remoter regions.

Labour costs

Non-wage outlays amount to as little as 22 per cent of total labour costs in Denmark and to an astonishing 102 per cent in Italy (*Economist*, 25 January 1992: 80). As a consequence, to compare labour costs in the countries of the EU one ought to examine total expenditure on labour (which includes pension contributions, national insurance, etc.) and not just the direct wage or salary costs. Unfortunately, the data available with which to examine trends over time seem to be confined to direct wage costs. As a result, the available evidence must be treated with some caution. Figure 5.2 provides a convenient datum for 1992, showing the breakdown of labour costs in manufacturing into wage and non-wage components (this split is not available for Spain). The difference in hourly labour cost between the lowest cost and highest cost EU country is roughly in the ratio of 1:4, and the

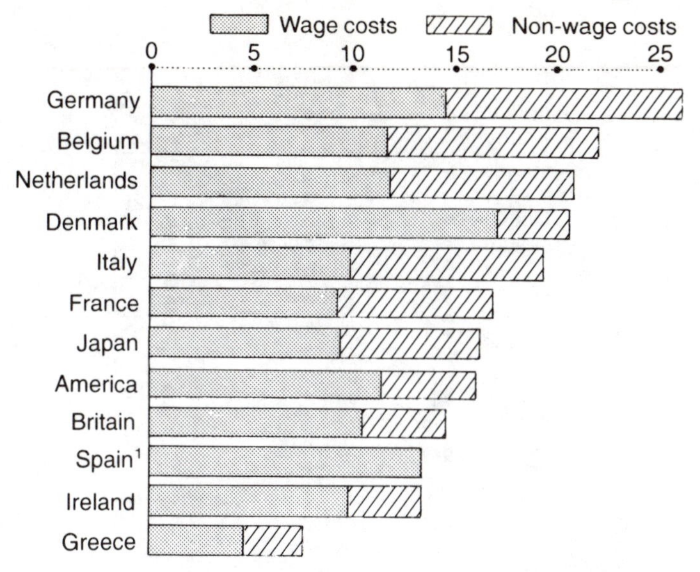

Figure 5.2 Labour costs in manufacturing, 1992 $US per hour
Note: [1] Split between wage costs and non-wage costs not available
Source: *Economist*, 3 July 1993

geographical pattern accords with one's expectation that high-income countries should have high labour costs in comparison with poorer countries.

Because, the data available for changes over time relate to wage costs only, and therefore ignore changes in the non-wage component, considerable care must be taken in examining the historical data; we have to assume that the ratio of wage to non-wage costs in each country has remained constant, an assumption which clearly is rather heroic. This caveat must be kept in mind in the following discussion.

Flanagan (1993) has examined the available data on wages for the EUR6 and EUR9 countries over the period 1957 to 1989. Until the late 1960s the geographical dispersion of wage rates remained either constant or declined, depending on the data series employed. From 1968 or 1969, wage disparities increased sharply for about a decade and then fell equally rapidly, returning in the late 1980s to the level experienced in the early 1960s. These temporal changes apply to both the EUR6 and the EUR9 countries and appear to have been unaffected by the timing of liberalization of intra-EU migration laws and the accession of three states in 1973.

Figure 5.3 shows the evolution of hourly wages (not total labour costs) between 1970 and 1991, with a forecast to 2010, for the EUR12 countries but excluding Luxembourg. Between 1970 and 1991, the pattern for eight of the eleven countries is one of clear convergence towards the EU average. The

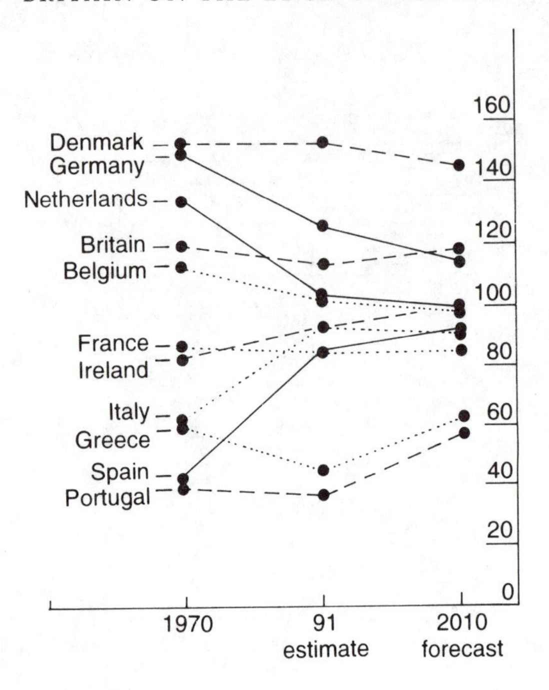

Figure 5.3 Hourly wages as percentage of EU average
Source: *Economist*, 25 January 1992

three countries in which relative wage costs stayed roughly constant – Denmark, Greece and Portugal – are all relatively small, so that overall the geographical variation in wage costs clearly declined quite sharply. If the ratio of wage to non-wage costs remained constant, then there was a similar convergence in hourly labour costs. The study cited by the *Economist* on which Figure 5.3 is based finds reason to believe that there will be some further narrowing of wage differences to 2010. The *Economist* itself (25 January 1992: 80) argued that two processes are likely to push total hourly labour costs further towards the EUR12 average. Multinational companies operating throughout the EU may become less willing to countenance the existing diversity of non-wage costs. In addition, to the extent that eleven out of the twelve countries implement the Social Chapter of the Maastricht Treaty, pressure will grow from governments to standardize non-wage labour payments.

For our purpose, it is more relevant to concentrate on the period 1970–91, in which there has already been a significant diminution of wage differences between the EUR12 countries, than speculating about the future. The historical evidence is more consistent with the neo-classical view that spatial processes are equilibrating in nature in the long run, than it is with the view that cumulative growth at the regional/international scale will persist over long periods of time and accentuate differences. In addition, this recent evidence suggests that the equalization of factor prices which Tovias (1982) found had been occurring even prior to 1958, and which had apparently come to an end in 1975, has in fact continued in response to the progressive integration of the European economies.

CONCLUSION

Several strands of evidence suggest that the economies of the EUR12 countries have been converging during the past thirty years. National differentials in GDP per person, unemployment levels and labour costs have all narrowed, with the clear implication that capital formation plus inward investment has been more active in the poorer countries than the richer. Migration has probably played a rather small part in these equalization processes.

The negative conclusion that we can draw from this evidence is that the prediction derived from cumulative growth theory – that the regions of highest economic potential will progressively outstrip the more peripheral regions – is not substantiated at the national scale of analysis. The positive conclusion to be drawn is that the equilibrating processes postulated by the neo-classical school of thought cannot be written off and may actually be working. Many of the countries with relatively low wages have been very successful in their development efforts, one consequence of which has been the reduction in labour-cost disparities.

During the 1980s, Keeble published several works framed around the hypothesis that the prosperity of a region is affected by its location on the map of economic potential. In more recent work, examining the evidence of what is happening to incomes, unemployment and industrial restructuring, he proposes that the core-periphery model is too simple (Keeble 1989, 1991a, 1991b). Summing up the evidence, he says:

> The complexity and variety of economic forces currently at work in Europe's regions are too great to be encompassed by any single all-embracing theory of economic change. The outcomes of these forces at the regional level are moreover not inevitable, but contingent upon many factors, not least social and political responses by local communities, institutions and governments. The result is a 'regional mosaic' of development trajectories within Europe, in the evolution of which both macro-economic forces and local socio-economic characteristics are important.
>
> (Keeble 1991b: 53)

Keeble is not alone in drawing attention to the complexity of trends within the EU and to the fact that a simple core-periphery model is inadequate (e.g., Begg and Mayes 1993; Gordon and Thirlwall 1989). The suggestion that the

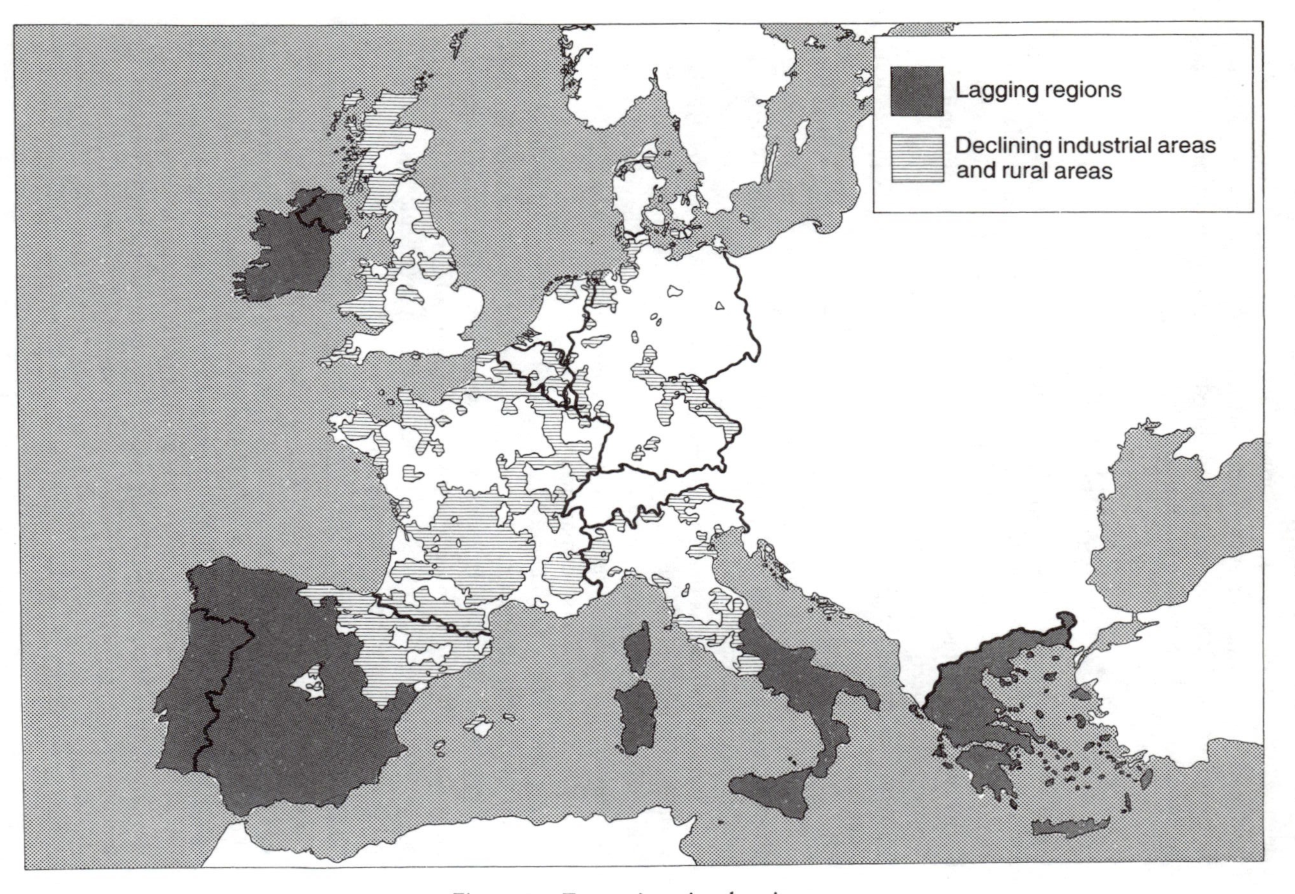

Figure 5.4 Europe's assisted regions
Source: Commission of the European Communities, 1991: 48

reality consists of a mosaic, rather than as a clear-cut core-periphery distinction, is amply confirmed by the map of areas eligible for regional assistance from EU funds (Figure 5.4). The low-income regions of Greece, Ireland, Portugal, southern Italy and Spain stand out as large blocks of territory on the periphery of the present-day EU. But scattered throughout the rest of EUR12 are large areas suffering from the problems of industrial restructuring, and many of these areas are truly within the core of the Community. These areas of industrial restructuring generally coincide with substantial levels of unemployment. Furthermore, whereas in 1993 the British government added areas in the south-east of England to the map of domestically assisted areas, the European Commission in March of the same year designated part of Wallonia in Belgium as an Objective 1 assisted region, thus putting it 'among the EC's poorest regions, on a par with Ireland, Greece, Portugal and Eastern Germany' (*Economist*, 17 April 1993: 49). A central location clearly provides no guarantee of prosperity. To reinforce the point, in both France and Spain there has been strong growth of the industrial base, well above the national averages, in many of the more peripheral regions, while at the national level the highest rates of growth in manufacturing output in the 1980s took place in Ireland and Portugal (Keeble 1991a, 1991b).

On the evidence presented in this chapter we can at least come to a firm negative conclusion that there is nothing in the recent history of the EUR12 countries to suggest that centralization processes necessarily dominate over processes which lead to a more dispersed pattern of economic activity, if we pitch the analysis at the national or macro-regional scale. This conclusion reinforces the conclusions reached in earlier chapters, on other evidence, that Britain's location on the edge of Europe is not an intrinsic handicap.

6

ACCESSIBILITY COSTS AND REGIONAL DEVELOPMENT IN BRITAIN

I'll put a girdle round about the earth
In forty minutes
Puck, in *A Midsummer-Night's Dream*

In Chapter 3 we explored the role of distance in international trade, and hence the significance of relative location in the process of international economic development. A limited amount of direct evidence on the cost of international transport was cited in that chapter. We now turn to examine the information which is available on the role of transport costs in regional development in Britain, again starting from the logic of classical location theory which posits the advantages of locations which minimize transport costs. This frame of thinking leads to the proposition that areas which are distant from the main centres of population, suffering from low economic potential, are at a permanent disadvantage compared with more accessible areas, being subject to higher transport costs than are more accessible regions. However, assuming for the moment that there are indeed measurably higher transport costs in the remoter regions, we also need to consider whether, and to what extent, these higher transport costs may be exacerbated or offset by variations in expenditure on other things, such as wages and land costs.

To initiate this discussion we will first examine the role of transport at the aggregate level, in particular the contribution of the transport sector to GDP, the structure of the transport sector in terms of the contribution made by the freight and passenger segments, and Census of Production evidence on transport costs. This discussion will set the scene for an examination of the extent to which location within Britain affects transport costs and hence the potential for economic development.

TRANSPORT IN THE NATIONAL ECONOMY

In the industrialized economies of the EU, the transport sector contributes between 4 and 8 per cent of GDP and Britain is no exception. The national

Table 6.1 United Kingdom: GDP at factor cost and current prices, percentage shares for transport and communications

	Transport plus communications		*Transport*	*Communications*
1971		8.3	5.9	2.4
1976		8.4	5.6	2.8
1981	7.4	7.3	4.7	2.6
1986	7.2	7.3	4.5	2.8
1987	7.2	7.4	4.6	2.8
1991	7.0			

Note: Since 1987, transport and communications have been combined. The combined series has been published back to 1978 but differs slightly from the sum of the two sectors available in earlier publications

Source: *United Kingdom National Accounts*

income accounts show that the transport sector now contributes about 4.5 per cent of GDP, whereas in 1971 the proportion was just under 6 per cent (Table 6.1). This table also shows that whereas the transport sector is declining in relative importance, the contribution of communications is growing in significance, and index numbers published in recent issues of the *United Kingdom National Accounts* show that this shift between the two sectors has continued since 1987.

These GDP estimates show the transport and communications sectors as being more important than is indicated by the 1984 industrial input–output table (*United Kingdom National Accounts* 1988: Table 2.6). According to this tabulation of transactions between producers of goods and services, purchases from the transport sector amounted to 2.4 per cent of total output, and from the communications sector just 1.1 per cent – in both cases approximately half the estimates derived from the GDP data. One of the reasons for this difference is the fact that an input–output matrix only takes account of purchases and therefore does not distinguish the transport and communications services which firms provide for themselves.

A third source of aggregate data is the Census of Production. These censuses used to be taken at irregular intervals and with rather variable coverage. More recently, annual surveys are undertaken but their utility in the present context is not great since transport (and communication) costs are not recorded. The published results for earlier censuses do not include a full tabulation of transport costs, since data for own account provision are incomplete, notwithstanding that own account services amounted to about 40 per cent of total expenditure. Nevertheless, one can estimate that total transport costs (purchased plus own account) amounted to the following percentages of gross output: 1963, 2.2; 1968, 2.4; 1980, 2.2.

All three sources of information are in agreement that the transport sector is a fairly small component of the national economy. Even so, it may be the

Table 6.2 Britain: growth of GDP and inland traffic (1985 = 100)

Year	*GDP at constant factor cost, 1985 prices UK*	*Freight*[1] *Tonne kilometres GB*	*Tonnes lifted GB*	*Haul kms GB*	*Passenger*[2] *kilometres GB*
1961	62.5	55.8	86.6	64.4	49.9
1966	72.4	66.5	108.0	61.6	64.3
1971	81.1	72.1	104.7	68.9	77.9
1976	87.0	79.2	101.8	77.8	74.4
1981	89.3	92.4	91.8	100.6	91.1
1986	103.8	100.0	101.7	98.2	104.3
1991	113.8	113.3	109.9	103.1	126.7

Notes: [1] Road, rail, coastal shipping and inland waterways
[2] Private vehicles; public transport by road, rail and air
Source: *Annual Abstract of Statistics*

case that, because transport costs should vary systematically in space, they will result in significant differences in the prosperity of different regions.

Table 6.2 summarizes the growth of GDP, freight traffic and passenger traffic over the period 1961–91, taking 1985 as the base year and equal to 100. Despite cyclical fluctuations in economic activity, GDP has grown fairly steadily over the thirty years to reach a level nearly double that of 1961. The volume of freight traffic, measured in tonne-kilometres, has increased fractionally faster. There was a sharp increase in the tonnes lifted in the early 1960s; however, after 1966 the total tonnage declined to 1981 and has only recently again reached the level of 1966. The fact that the tonnage of goods shifted within Britain has not increased in the last twenty-five years need occasion little surprise. It is well known that technological advances are continuously giving rise to greater efficiency in the use of materials and fuels, with the result that the physical volume of resources needed to produce each unit of GDP has been declining (Ihde 1991). This trend has been accentuated by structural shifts in the economy away from 'heavy' industry (such as steel) and towards much higher value added products, including electronic information systems hardware. In addition, the manufacturing sector as a whole has lost ground relative to the service trades of all kinds, from the financial sector to tourism; these activities require relatively small inputs of commodities and large contributions of manpower and capital. These overall shifts have had an obvious impact on the tonnes lifted, and go a long way to accounting for the limited growth that we have noted. In addition, electricity has had a substantial impact. On the one hand, conversion of primary coal, oil or gas, and transmission of the resulting power through the national grid, removes an element from the category of data conventionally regarded as the transport sector, even though the reality is that one form of transport has been substituted for another. Furthermore, the development of the nuclear generation programme to the point at which it accounts for about one-fifth of all electricity produced implies a sharp drop in the quantity of primary fuel that must be moved, reinforcing the decline occasioned by the shift from coal, which has a relatively low calorific value, to oil and gas.

While such considerations account for the lack of growth in the tonnes lifted, they do not explain the substantial fluctuation thereof. Part of the explanation may lie in the volume of construction business that is in hand at any one time, since the construction industry is a big user of ballast, cement, bricks and other bulky goods, and is also susceptible to larger oscillations in business activity than is the economy as a whole (Department of Transport 1993: 3).

With respect to freight, the two most striking features of Table 6.2 are the increase in mean haul and tonne-kilometres of traffic. In 1961, the average distance over which goods were moved was 67 kilometres. Between 1961 and 1965, the mean fell to the lowest value of 63 kilometres, since when it rose steadily to a maximum of 107 kilometres in 1984, a figure which was only again reached in 1991. This 60 per cent increase in mean haul is consistent with the steady improvement of the road system, with rapid technical and other improvements in transport systems leading to a reduction in the real cost of transport, and to the shift in the composition of traffic towards goods with higher unit values, to mention just some of the relevant factors. The net effect has been an extension of the radius of interaction.

Personal mobility has increased dramatically. Passenger kilometres have increased by 150 per cent, from about 270 billion in 1961 to about 690 billion in 1991. This is a rate of increase far in excess of the expansion of GDP and of freight traffic. Note that passenger traffic includes both business and non-business journeys.

At the aggregate level, there have clearly been important changes in the scale and nature of the transport sector within the national economy. The fact that the volume in traffic has expanded more rapidly than total national product, while the transport sector has been contributing a declining proportion of GDP, is clear proof that the transport sector's productivity has been rising more rapidly than productivity in the economy as a whole, making transport progressively cheaper in real terms.

TRANSPORT COSTS AND THE SPACE ECONOMY

Measurement of the role of transport in the space economy may be undertaken in one of two ways; by means of *ex ante* or of *ex post* analysis. *Ex ante* studies approach the problem by asking the question: what would the transport costs be for firms were they to choose particular locations in the set of feasible locations? To answer this question, one must specify the nature of a firm's activities and make an assessment of both its transport needs and the cost of providing for those needs. This approach is very similar to the way in which a firm itself would review its investment options. In contrast, *ex post* studies enquire into the actual experience of existing firms in different locations, to establish what transport costs have been in practice. Both approaches have their advantages and disadvantages. However, they share an equal but opposite problem. In the *ex ante* approach there is no certainty, if a firm chose one of the locations identified, that its actual experience there would accord with prior expectations. Conversely, in an *ex post* study it is dangerous to assume that if other firms chose to locate in a particular place or region they would experience transport costs similar to the firms already located there, for which data have been collected. Therefore, neither form of analysis on its own can give a completely unequivocal answer to the question whether locational differences do have a measurable impact on costs and hence profits.

Most of the available studies are *ex post* in character. Fortunately, the testimony derived from these studies is consistent with the evidence available from the limited number of *ex ante* analyses, so that we are spared the problem of reconciling major conflicts in the conclusions reached. As McKinnon (1992: 46) observes of the Scottish economy in relation to the rest of Britain, 'Available evidence suggests that, in the case of Scotland, peripherality neither imposes a significant transport cost burden on manufacturers nor induces them to make sub-optional trade-offs between transport and inventory.'

EX POST STUDIES

We will begin our review of the evidence which lies behind McKinnon's conclusion by first considering the *ex post* analyses, of which the Toothill Report (Toothill 1962) was the first major post-war study of immediate relevance. Forty-five firms with experience of operating in Scotland and also south of the Border were asked the extent to which transport costs were higher in Scotland. As Table 6.3 shows, substantially more than half of the respondents estimated that the extra transport costs added no more than 1 per cent to their operating costs; a further quarter put the increase at between

Table 6.3 Increase in firms' operating costs on account of transport, Scotland compared with southern Britain

	1% and under	*1–3%*	*3–5%*	*Over 5%*
Number of firms	26	11	4	4

Source: Toothill 1962: 74

1 and 3 per cent. These figures do suggest that there is a disadvantage for firms located in Scotland, but that it is small. However, the Toothill Committee found that these higher transport costs were fully offset by lower costs on other inputs, such as land and labour:

> In short, we found nothing in our enquiries to support the view that transport costs are a significant additional burden on manufacturing industry in industrial Scotland. Indeed we have seen no evidence to suggest that over a wide range of industry we cannot produce as cheaply in Scotland as anywhere else in Britain.
>
> (Toothill 1962: 75)

This conclusion quickly found support in the major study of regional development in Britain by Brown (1969, 1972). He undertook a thorough review of the extant evidence pertaining to spatial variations in costs and profitability, including the Toothill Inquiry, and came to the following conclusions:

> Broadly speaking, the trend of thought in this field has been towards the realisation that transport costs are of relatively minor importance in the majority – and an increasing majority – of industries, that adequate supplies of trainable labour (for some purposes, and for some firms, already-trained labour) are of paramount importance in the post-war situation of relatively full employment, that managerial communications with clients, suppliers, sub-contractors, colleagues and various professional services loom large, and that amenities are important – these last two especially to the people who make the locational decisions.
>
> (Brown 1969: 778)

Other *ex post* evidence appears to contradict the conclusions reported above. Glasgow University conducted a survey of firms which had established branch plants in Assisted Areas in the period 1958 to early 1964. A total of 142 such companies were identified, and responses were obtained from eighty. Of the respondents, twenty-five had established plants in Scotland (Cameron and Clark 1966; Cameron and Reid 1966). In the case of eighteen plants, a Scottish location had been considered but explicitly rejected in favour of somewhere else. Enquiry of these eighteen firms indicated that anxieties about access to customers – not suppliers – was an important consideration in the minds of decision-makers. On the basis of this limited evidence, Cameron and Reid questioned the reliability of the Toothill findings. These doubts were subsequently reinforced by examination of the 1963 Census of Production. Transport cost as a proportion of net output was somewhere between 11 and 30 per cent higher in Scotland than the national mean, with East Anglia, the North and Wales as the other notably high-cost regions. Overall, the range from highest- to lowest-cost region was between

Table 6.4 Transport costs in Scotland compared with the United Kingdom, 1963 and 1974: purchased transport plus own-account expenditure

1963

Industry, 1958 SIC	*Total transport cost as % net output, larger firms only* — *Scotland*	*UK*
iii Food, drink, tobacco	11.5	13.9
iv Chemicals and allied	5.5	7.0
v Metal manufacture	7.4	6.7
vi Engineering and electrical goods	2.2	2.7
vii Shipbuilding and marine engineering	1.2	1.2
viii Vehicles	1.7	2.2
ix Metal goods n.e.s.[1]	6.2	5.1
x Textiles	3.1	3.3
xi Leather, leather goods, fur	6.0	4.6
xii Clothing and footwear	2.4	2.5
xiii Bricks, pottery, glass, cement, etc.	17.8	15.6
xiv Timber, furniture, etc.	10.5	9.3
xv Paper, printing, publishing	7.3	6.0
xvi Other manufacturing	6.3	4.5
Total	5.9	5.8

1974

Industry, 1968 SIC	*Total transport cost as % gross value added, all firms* — *Scotland*	*UK*
iii Food, drink tobacco	11.8	16.0
iv and v Coal and petroleum products, chemicals and allied industries	5.6	5.7
vi Metal manufacture	6.6	6.6
vii Mechanical engineering	3.3	4.1
viii Instrument engineering	1.2	3.6
ix Electrical engineering	3.6	3.8
x Shipbuilding and marine engineering	1.2	1.7
xi Vehicles	4.1	3.7
xii Metal goods n.e.s.[1]	8.8	6.2
xiii Textiles	3.6	4.6
xiv Leather, leather goods, fur	5.7	6.0
xv Clothing and footwear	3.1	4.2
xvi Bricks, pottery, glass, cement, etc.	16.0	14.7
xvii Timber, furniture, etc.	12.0	10.0
xviii Paper, printing, publishing	7.1	6.7
xix Other manufacturing	7.8	6.0
Total	6.755	6.827

Note: [1] n.e.s. = 'not elsewhere specified'
Source: Chisholm 1987: 307, 309

2 and 3 per cent of net output (Chisholm and O'Sullivan 1973; Edwards 1975). However, these estimates were based solely on the cost of purchased transport and therefore excluded expenditure on transport services provided by firms for themselves, at that time amounting to 40 per cent of total expenditure. It seemed reasonable to assume that the outlays on purchased and on own account transport would be closely correlated at the regional level. However, that assumption has turned out to be erroneous.

Building on the earlier work of Logan (1971) and the Scottish Office (1981), it has been possible to analyse further the Census of Production data for 1963 and 1974 to examine purchased and own account transport costs in Scotland compared with the rest of the country. In both census years the cost of transport as a percentage of net output (or gross value added) in Scotland was almost identical to that for the United Kingdom (Table 6.4). If costs are calculated by standardizing for industrial structure, the same picture emerges. Therefore, the overall similarity of Scotland and the United Kingdom reflects the industry-by-industry similarity of transport costs, and not the chance of industrial composition (Chisholm 1987), with the implication that Scotland suffers no transport cost disadvantage on account of its peripheral location.

A much-cited study by PEIDA (1984) has been used to argue that, notwithstanding the evidence cited above, peripheral areas are at a significant disadvantage on account of measurable and unmeasurable costs of access. The PEIDA study obtained evidence from 515 firms located in Northern Ireland, Scotland and South East England. From this survey, the result which has attracted the most attention is the *perception* of firms in Northern Ireland and Scotland that they suffer significant disadvantages on account of access problems (e.g., Keeble *et al.* 1986: 11–12). In terms of the costs that could be *measured*, however, the penalty of peripheral location proved to be quite small for the firms sampled. Measured transport cost as a proportion of gross value added in manufacturing was estimated to be: Northern Ireland, 10.8 per cent; Scotland, 8.3 per cent; South East England, 7.7 per cent. The concluding text which accompanies these figures says:

> Regional differences were not great, indeed the interview sample did not produce a clear relationship between peripherality and transport costs – this may reflect marketing patterns among the firms surveyed. The postal survey with its larger numbers and wider scope found transport costs to be marginally higher in Scotland than in South East England, with a rather more substantial difference between South East England and Northern Ireland.... The measured transport cost differences found – certainly between Scotland and South East England – do not seem capable of accounting for observed differences in regional economic performance. In Northern Ireland, the cost differential is more significant, though still, in itself, unlikely to account for the weakness of the region's economic performance.
>
> (PEIDA 1984: 87)

The PEIDA report attempts to reconcile the paradox that although the firms in the two peripheral regions felt that they were at a disadvantage on account of higher transport costs, this perception was not translated into significant measurable differences. It is suggested that the managers of firms in peripheral locations devote more time and energy to transport matters than do their counterparts in more central locations. And it may be that although the measurable costs of transport do not vary greatly, there are unmeasured (and perhaps unmeasurable) aspects of transport which result in greater disadvantage in peripheral areas than elsewhere.

EX ANTE STUDIES

The whole *ex post* approach may be criticized on several grounds, of which two are particularly important. First, for the Census of Production – but not, of course, for special surveys of firms – it is becoming increasingly common for a firm which has several plants to make a single, combined, return for all of them. Over time, therefore, the Census is becoming less reliable for spatial analyses. Second, in obtaining data from many firms in various locations, all *ex post* studies fail to compare like with like. Therefore, one can argue that the only way to obtain safe conclusions is to develop an *ex ante* model, to show what the position would be were a given firm to elect one location in preference to others (Tyler and Kitson 1987; Tyler *et al.* 1988a, 1988b).

Two studies are available which show spatial variations in transport costs on an *ex ante* basis. Although they both show that Scotland and Wales experience higher transport costs, the magnitude of the spatial variation differs markedly in the two cases. McKinnon (1992) reports an earlier study by Tarry which was structured in the following manner. Tarry took Central Scotland and seven cities located in England and Wales. For each of these eight origins, he obtained quotations from a sample of seventeen road hauliers for delivery to each of nine destinations in mainland Europe. The mean of the seventeen quotations was taken to represent the cost of transport on each of the seventy-two routes (eight origins × nine destinations). If we assume that a firm located in each of the eight origins shipped the same volume of goods to each of the nine destinations, the total transport cost for exports may be calculated. The results obtained from this exercise are shown in Table 6.5. With the exception of Swindon, the pattern conforms to the expectation that the north and west of the country should experience higher transport costs, but the range is quite small; the two most favoured locations, Peterborough and Telford, have an average transport cost to the nine European cities 15 per cent below the cost for Central Scotland.

Table 6.5 Average transport costs by road from eight UK locations to nine European cities (Central Scotland = 100)

Central Scotland	100	The nine European cities:
Swansea	95	Amsterdam
Newcastle	93	Berlin
Liverpool	91	Brussels
Middlesbrough	91	Frankfurt
Swindon	91	Helsinki
Peterborough	85	Milan
Telford	85	Munich
		Paris
		Stockholm

Source: Tarry, quoted by McKinnon 1992: 32

However, that result is at variance with another *ex ante* study, which shows a much larger geographical variation in transport costs. Tyler and Kitson (1987) divided Great Britain into thirty regions and considered the spatial variation in transport costs facing representative but hypothetical firms in delivering their output to customers. Two situations of particular relevance in the present context were envisaged: that of firms selling partly on the domestic market and abroad; and that of firms orientated to exports. Some heroic assumptions were made concerning the geographical distribution of outputs, namely:

1 Domestic final consumption is distributed in proportion to population.
2 The market for intermediate goods is distributed in proportion to the level of 'economic activity'.
3 Export consignments use all ports in proportion to the volume of traffic which they handle.

The effect of these assumptions is to define the geography of markets in a manner that is invariant with the location of a firm in any one of the thirty areas. Information was then obtained concerning transport costs within Britain. Armed with these data and assumptions, the authors found that the lowest inland transport costs for an industry selling partly on the home market and partly overseas (e.g., mechanical engineering) would be in Birmingham; for an export-orientated industry, the minimum-transport-cost location would be London. In both cases, the highest-cost location would be Inverness, at a little under three times the cost of the optimal locations. This ratio of nearly 3:1 is considerably greater than that obtained by Tarry, whose data included both the inland and overseas transport costs. Clearly the specification of the *ex ante* model is liable to have a significant effect on the results obtained.

However, let us for the moment assume that the results obtained by Tyler and Kitson are reasonably representative of the real situation. A companion study by Tyler *et al.* (1988a, 1988b) shows pretty conclusively that geographical variations in transport costs of the order of even two to one are not material in influencing the level of profitability of firms. The approach is to imagine a firm contemplating an investment which might be located in any one of the English counties. To solve this problem, two sets of data are needed: the structure of operating costs for the firm, and the geographical variation in those costs. The Census of Production identifies the main cost headings for whole industries, in terms of expenditure on wages, fuel, purchased components, transport costs, etc. If we assume that the national cost structure of an industry represents the structure for a representative firm in that industry, the Census provides the data to give the structure of operating costs for firms in the range of industries identified for census purposes. The second, and more laborious, task is to obtain data on the

Table 6.6 England: mean contribution of cost components to deviations in gross profits from Inner London

Cost component	*Percentage contribution to higher profits*
Salaries	+50.8
Industrial services	+24.1
Other (energy, bought in services)	+ 8.5
Rates	+ 8.3
Wages	+ 5.9
Rent	+ 2.8

Note: The figures are the unweighted means for the contribution to variations in gross profits with respect to Inner London in twenty counties plus Outer London, at distances up to 400 kilometres from London. They do not add exactly to 100
Source: Tyler *et al.* 1988a: 41

geographical variation of these costs, armed with which it is possible to see how far costs and profits would vary from one location to another for firms with given cost structures. For this purpose, it was assumed that the geographical pattern of sales would not vary with the firm's location choice.

This procedure was applied to about one-third of the 101 minimum-list manufacturing industries identified within the 1980 Census of Production, these industries being selected as representative of manufacturing as a whole. In every case, Inner London proves to be the least profitable location. Furthermore, for all of the industries, profits rise steadily with increasing distance from London, such that 'almost 70% of industries would experience a profits increase of more than 30% by moving 200 miles [320 kilometres] into the South West peninsula' (Tyler *et al.* 1988a: 39–42). This finding is contrary to the expectation, based on transport costs alone, that peripheral regions are at a disadvantage in respect of costs, and hence profits. The primary reason that profits, in this formulation, rise with distance from London is the fact that, although transport costs are higher, this locational penalty is more than offset by lower costs for virtually every other item of expenditure.

Table 6.6 summarizes the role of the various factors in contributing to higher operational profits away from Inner London. For each of the factors, the contribution has been obtained as the unweighted mean of the estimates made for the twenty counties plus Outer London. It is quite clear that, on the assumptions made by Tyler *et al.*, geographical variations in profitability are dominated by salaries and industrial services, both of which are cheaper away from London, but this benefit is partially offset by the estimated increase in transport costs, although this effect is masked by the inclusion of transport in the category 'other'.

It is in fact hardly surprising that transport costs should play such a small part in geographical variations in profits as measured in this *ex ante* fashion. In 1980, gross profit accounted for 10.5 per cent of the value of gross output in manufacturing industry in the United Kingdom, the other 89.5 per cent being attributed to expenditure of all kinds by firms – on materials, wages, transport, etc. In 1980 the value of *purchased* transport, as recorded by the Census, was 1.3 per cent of gross output (if own account transport is added in, the total transport cost rises to about 2.2 per cent). Clearly, however big the spatial variations in transport costs may be, the potential for affecting profit levels is very small in comparison with the potential impact of spatial variations in other costs. The potential impact of transport is in any case reduced if we recollect that Tyler *et al.* were working with data on purchased transport costs, whereas in 1963 and 1974, and probably also in 1980, spatial variations in own account transport expenditure tended to offset the spatial variability of purchased transport costs.

If everything else were equal, then variations in transport costs would be the determinant of regional variations in profits and prosperity. In practice, spatial variation in transport costs is only one among many factors which may vary systematically in space, and the question to be addressed is the relative significance of transport compared with other factors. *Ex post* studies show that in Britain regional differences in transport costs are small or negligible, whereas *ex ante* models suggest that there is somewhat greater variation. However, both approaches agree that if there is a systematic variation in transport costs this is compensated by variation in the cost of other factors. Within the United Kingdom, the peripheral regions are not in fact disadvantaged by their remoteness. The fact that both the *ex ante* and the *ex post* approaches reach the same basic conclusion suggests that it is indeed robust.

CHARACTERISTICS OF THE FREIGHT HAULAGE INDUSTRY

To explain why the *a priori* expectation that there should be substantial regional differences in transport costs is not borne out in practice, we need to examine some of the characteristics of the freight haulage industry. In the first place, for goods to be moved at all, they must be loaded, unloaded, recorded and invoiced. These fixed costs are considerable in relation to the cost of movement. For example, PEIDA (1984: 18–19) report a 1983 study of distribution costs in British industry. The sum of storage costs, stock interest and administration amounted to 8.6 per cent of the value of sales, compared with transport costs of 3.7 per cent. Take this in conjunction with the relatively short distance over which freight moves, and we have a major reason for the smallness of geographical transport cost variations. In the 1960s, the average haul for all freight was in the order of 60–65 kilometres, and for road traffic about 50 kilometres. Data for 1966 showed that in the

Table 6.7 Great Britain, 1992: percentage distribution of road freight by length of haul

	Length of haul in kilometres							
	Up to 25	*25.1–50*	*50.1–100*	*100.1–150*	*150.1–200*	*200.1–300*	*Over 300*	*Million tonnes*
Tonnes								
Food, drink, tobacco	17	20	23	14	8	10	8	290
Bulk products	50	23	14	5	3	3	2	570
Chemicals, petrol, fertilizer	18	21	25	13	7	8	8	118
Miscellaneous products	42	15	12	8	6	9	8	486
All commodities	39	19	16	8	5	7	6	1,463
Tonne-kilometres								*Million tonne-kilometres*
Food, drink, tobacco	2	6	15	15	13	22	27	33,156
Bulk products	10	13	17	12	10	17	21	36,498
Chemicals, petrol, fertilizer	2	7	17	14	10	19	31	12,700
Miscellaneous products	6	6	10	11	12	22	33	38,895
All commodities	6	9	14	13	11	20	27	121,250

Source: Department of Transport 1993: 31–2

road haulage industry goods must be moved about 125 kilometres for the movement costs to equal the fixed costs. Taking the two points – short average hauls and high fixed charges relative to movement charge – it appears that in the 1960s about two-thirds of national freight costs were unaffected by the length of haul and hence inter-regional location (Chisholm 1971; Chisholm and O'Sullivan 1973). A rather similar result had been recently obtained by Bayliss and Edwards (1970) on the basis of a sample of consignments. This conclusion may seem surprising but in the two decades since the findings were published I have not been aware of any study which challenges either the data or the assumptions used in the analyses just reported.

Furthermore, although the mean length of haul has increased since the 1960s, the great bulk of freight traffic remains essentially local. Of all freight moved in Great Britain in 1982, 77 per cent stayed within the standard region of origin; 62 per cent in the case of rail traffic, and 78 per cent in the case of goods moved by road (Chisholm 1985b: 968; *Regional Trends* 1984: 131–2). Ten years later, the proportion of road freight which moved intra-regionally had fallen somewhat, to 72 per cent, or still almost three-quarters of the total. As Table 6.7 shows, almost 40 per cent of traffic by tonnage moves no more than 25 kilometres, and three-quarters not more than 100. The implication is that a very substantial part of transport costs must remain unaffected by inter-regional location, notwithstanding that over half the tonne-kilometres of (road) freight traffic is recorded on hauls of over 100 kilometres. That implication is confirmed by the responses of firms interviewed by PEIDA in Northern Ireland, Scotland and South East England: 'Among the majority of firms interviewed only round 50% of freight charges for deliveries within the UK were accounted for by the distance related element' (PEIDA 1984: 80). This finding is consistent with the evidence in Table 6.5, showing a comparatively small spatial variation in quoted freight charges from a number of locations in Britain to a specified set of continental destinations, and with the other evidence which has already been cited for earlier years.

In addition to the fact that freight charges contain a significant element of fixed costs and are therefore not proportional to distance, firms supplying goods often choose to quote the same delivered price to customers irrespective of their location. Uniform delivered prices represent the extreme situation, in which the purchaser is indifferent to relative location. Freight absorption is a common practice which reflects the fact that transport costs are relatively small compared with other costs (Chisholm 1966). For some bulky and lower-valued commodities, zonal rates are applied, the effect of which is to reduce but not eliminate the distance effect. To the extent that uniform or zonal rates are the norm, the incentive is for the supplier to be near his customers. However, since customers may be in several or many locations, and are apt to change over time, the pressure for agglomeration to occur is not great. Yet again, and relevant for container traffic in the export trades, is the basing point system. A container originating in Scotland and shipped through Felixstowe, for example, will be charged as if it were routed by the nearest Scottish port, say Greenock or Grangemouth. Finally, the charge which is made by a haulier is affected by the balance of traffic. In general, the volume of imports into Britain is such that there exists surplus capacity for handling exports, with the result that on some routes very favourable rates can be obtained (Chisholm 1966, 1987; McKinnon 1992).

All of the above considerations reduce the links between transport costs and distance. Taken together, and given that transport costs are in any case only a small part of industrial costs, it is not surprising that in practice the regional differentials in transport costs are small or non-existent and in any case are dwarfed by variations in other costs. This conclusion, based on attempts at direct measurement of differences within Britain, is clearly consistent with the situation described in Chapter 3 concerning international trade and the limited role that distance plays at that geographical scale.

RESPONSES BY FIRMS

Firms are naturally interested in minimizing their costs and maximizing their profits, within the constraints that they recognize, such as the maintenance of market share. It is reasonable to suppose that they will adopt practices that, given the circumstances under which they operate, will mitigate such problems as they encounter. Adaptive strategies will include efforts to reduce transport costs if these are perceived to be a problem.

We have already seen that, in the case of Scotland, adaptation in the form of self-selection of industries towards those which have low transport costs compared with the average of all industries has not occurred. Another adjustment which it seems can be discounted is the holding of larger stocks by firms in peripheral locations than by firms in more accessible locations. PEIDA (1984: 69–70) found that output stocks do not seem to vary regionally, although there is more of a tendency for input stocks to be larger in the remoter areas. However, this tendency was attributed not to the cost of transport but to unreliability in deliveries. In the absence of adjustments by firms in either of these two ways, the main means for minimizing any perceived or real transport cost disadvantage seem to be one or more of the following.

Firms in peripheral regions might adapt by obtaining a larger proportion of their inputs locally than is the case in more accessible regions. One would then expect that peripheral regional economies would be more self-sufficient than regions occupying more accessible locations. To some extent these expectations are borne out in practice. In addition, one would expect that the remoter regions would avoid any transport penalty arising from intra-national transfer costs by concentrating on overseas business to a greater extent. Again, this expectation is borne out to some extent (Chisholm 1985b;

McKinnon 1992; PEIDA 1984). To the extent that avoidance strategies are used by firms in this way, the savings in transport costs may be bought at the expense of quality or price competitiveness of purchased inputs and/or limited sales potential. However, given the small absolute level of transport costs and the limited degree of spatial variation thereof, the effect is likely to be small. Finally, firms may adapt by their marketing methods, and in particular the proportion of their output that they sell through wholesale agents rather than direct to customers. One hypothesizes that in remote locations a larger proportion of sales would be handled by wholesale agents. This appears to be the case for Northern Ireland, but not for Scotland, in comparison with South East England (PEIDA 1984, 57–8). This rather equivocal finding suggests that the nature of the industrial products may be as important as regional location.

Responses by firms of the kind mentioned above do call in question the wisdom of making *ex ante* assumptions which rule out the possibility that a firm's operations will adjust to local circumstances. That being the case, it is as inappropriate to rely exclusively on the *ex ante* approach as it is to place sole reliance on *ex post* analyses. All the more reason, one may conclude, to be grateful that both approaches yield very similar results.

OTHER RELEVANT EVIDENCE

An alternative approach to the role of location in regional development is provided by attempts to measure growth and productivity changes at the regional level, and to establish the reasons for the observed patterns. The most thorough recent study of Northern Ireland compares manufacturing productivity in that region with the rest of the United Kingdom and with Ireland, Germany and the United States (Hitchens *et al.* 1990, 1991, 1992). By any of the accepted measures, both Ireland and Northern Ireland are peripheral to the main centres of economic activity in Europe, and both are generally perceived to be at a disadvantage as a result; this point is explicitly acknowledged in the Hitchens study. The key facts to emerge from this study are as follows. The productivity of manufacturing in Northern Ireland is lower than in mainland Britain, a feature of Ulster's economy which has persisted for a considerable time. On the other hand, whereas in the early 1960s productivity in Ireland and Northern Ireland was very similar, productivity levels in the Republic have forged ahead since then. In 1963, manufacturing productivity in the Republic was only 79 per cent of the United Kingdom's; by 1984, it had reached 127 per cent. This remarkable improvement was general to all manufacturing sectors and proves that the real or imagined handicap of a peripheral location can be overcome. In contrast, productivity has remained relatively low in Northern Ireland. In searching for an explanation, Hitchens and his colleagues draw attention to problems concerning capital investment, entrepreneurship, and with practices and attitudes to work, none of which is very obviously or directly related to relative location; in addition, political unrest may well be a material factor. Problems arising from the cost and/or quality of transport are accorded almost no attention as factors contributing to the poor performance of the Province's economy.

More generally, there has been a substantial literature on the 'North–South' divide in Britain, the Assisted Areas, the geography of unemployment, migration and incomes (e.g., Harrison and Hart 1993; Martin 1988, 1993; Townroe and Martin 1992). The conventional approach assumes that the peripheral regions are at a permanent disadvantage compared with the more favoured Midlands and South East. However, the problem has been to determine the direction of causation for the differences in relative prosperity. In the first place, it is quite clear that many of the northern and western regions had inherited rather specialized economic structures at the end of the Second World War, with the misfortune that these specialisms were in industries with poor long-term prospects so far as employment was concerned. Coal-mining, heavy steel, shipbuilding and cotton textiles are all examples of industries that formed the evaporating core of regional economies. The adjustment to an employment structure better focused on expanding sectors of employment has been a slow and painful process, and a good deal slower than most commentators and policy-makers had hoped and expected. There are signs that at last this transition has run the greater part of its course, as the recession of the early 1990s bit more deeply in southern England than elsewhere.

It has also become apparent that the urban–rural distinction is an important facet of regional dynamics. The first major study to point this out was published in 1982 (Fothergill and Gudgin 1982). Since then, and particularly with the publication of the 1981 and 1991 population censuses, it has become increasingly clear that the big conurbations, London included, have been struggling to maintain their number of both residents and jobs. In contrast, smaller towns and many rural areas in quite remote locations have experienced remarkable growth, not just of the elderly retired and/or long-distance commuters, but also of people who work locally (Gould and Keeble 1984; Healey and Ilbery 1985; Keeble 1984; Keeble *et al.* 1992). In substantial measure, therefore, regional prosperity depends on the mix of large urban settlements and smaller, more rural towns and villages. Herein lies another adverse legacy for areas like South Wales and the North East, in that the proportion of the population living in sizeable towns and cities is relatively large. Such peripheral regions face a second period of adjustment, which overlaps with the adjustment of employment structure to which we have already adverted, and which will again clearly take many years to work through.

Neither of these adjustment problems arises specifically from the relative location of the various regions. Rather, they are the legacy of earlier development, much of which was based on the existence of natural resources at a time when transport costs were much higher in real terms than is now the case. It is in this context that the recent study by Harris (1989), covering the period 1963 to 1985, is of considerable interest. The focus of his study is an enquiry into the nature of the economic weaknesses which have beset the peripheral regions. The central conclusion is that for too long central government has concentrated on encouraging inward investment into the Assisted Areas and paid too little attention to fostering indigenous firms, with the result that a high proportion of employment is in branch plants; in the case of Devon and Cornwall, external investment controlled 64 per cent of manufacturing employment in 1989–90 (Potter 1993). Although branch plants bring the benefit of immediate employment to areas sorely in need of work, the downside of a branch plant economy is the external control, and the associated lack of employment in R&D and the more entrepreneurial roles of management. However, since about the mid-1980s, which marks the terminal point of the Harris study, there have been indications that multi-plant companies are modifying their management structures to give individual plants greater autonomy. To the extent that this happens, the disadvantages of having large numbers of branch plants will be reduced, perhaps substantially.

Of main importance, however, is the fact that Harris, in considering the regions of the United Kingdom, confirms the Hitchens finding in respect of Northern Ireland, that accessibility problems really do not seem to be a major, or even a particularly significant, factor in explaining relative rates of growth. Since Northern Ireland is the least accessible of the peripheral

Table 6.8 United Kingdom: regional GDP per person as a percentage of the national average

	1966	*1971*	*1976*	*1981*	*1986*	*1991*
United Kingdom	100.0	100.0	100.0	100.0	100.0	100.0
North Yorkshire and	83.4	87.2	96.1	94.0	87.5	89.1
Humberside	96.0	93.0	94.8	91.8	93.5	91.6
East Midlands	97.2	95.9	96.4	96.3	96.9	98.1
East Anglia	94.6	94.0	94.1	95.2	101.0	100.4
South East	115.9	114.0	112.5	116.6	117.1	116.5
South West	90.8	93.0	92.3	95.6	94.5	94.2
West Midlands	106.7	103.0	97.8	89.8	91.6	92.0
North West	95.3	96.8	95.9	94.6	93.3	92.7
Wales	87.4	88.2	89.8	84.4	83.6	86.5
Scotland	89.1	92.3	98.0	97.8	94.7	96.7
Northern Ireland	64.9	74.3	74.7	75.6	80.2	77.1

Note: Continental shelf excluded
Source: *Regional Trends*

regions, the implication is that the other peripheral regions cannot be seriously disadvantaged by their relative location. In the long run, the relative prosperity of the regions will be determined by their comparative productivity, a point underlined by the fact that the relatively recently established Nissan plant near Sunderland, north-east England, in 1993 became Britain's biggest exporter of cars (*Independent*, 24 December 1993). The single most useful index of the relative standing of regions, and of changes over time, is regional GDP per person, and Table 6.8 summarizes the situation for each region, taking the United Kingdom average as equal to 100. The South East stands out as the region which has consistently had above average output per person; over the twenty-five years shown, there was only a very small increase in the relative advantage of this region. Two other regions were also stable in their relative standing, the East Midlands and Wales. The North West, Yorkshire and Humberside and the West Midlands all lost ground, the latter in spectacular fashion. The West Midlands used to be the only region other than the South East to have an above average level of GDP per head but now suffers the ignominy of being one of the poorer regions. The remaining five regions have improved their relative standing, especially East Anglia and Northern Ireland. Of these five regions, four are clearly peripheral – the North, Northern Ireland, Scotland and the South West. Taking these data as a whole, there is nothing to suggest that the peripheral regions must always be at a disadvantage relative to the regions which are more centrally located.

GRAVITY MODELS OF FREIGHT FLOW

The evidence which has been cited so far in this chapter all points to the conclusion that transport cost is of small importance as a factor explaining regional differences in profitability and hence regional prosperity. Yet there is evidence that the geography of freight transactions within Britain does conform to gravity-model formulations, and the conclusion which is generally drawn from these studies is that remoter and more peripheral areas are at a significant economic disadvantage. There would thus appear to be a conflict of testimony which needs to be addressed.

In Britain, the first gravity-model studies of freight flows were undertaken by the Ministry of Transport in the 1960s, including the influential analysis of the traffic potential of the new dock complex then proposed at Portbury, Bristol (Ministry of Transport 1966; see also Bassett and Hoare 1984). Shortly after, Chisholm and O'Sullivan (1973) reported the results of their more general study covering the whole of Great Britain, using 78 × 78 origin-destination matrices. This study itself reflected the widespread interest that there then was in gravity modelling (e.g. Black 1972; Wilson 1967, 1971) and was followed by further work on British data (Gordon 1976, 1978, 1985).

As with some of the studies considered in Chapter 3, the work mentioned in the previous paragraph does not explicitly record the proportion of freight flows between origin and destination regions which is attributable respectively to their size (or economic mass) and the intervening distance, and it is not now possible to estimate these effects from the published results. Much of the work concentrated on two issues: the goodness of fit that could be obtained with gravity models; and spatial variations in the distance exponent. For example, Chisholm and O'Sullivan obtained R^2 values in the range 0.73 to 0.80, and distance exponents for freight moved by road ranging from –1.3 to –4.8. The fact that this and other studies obtained negative distance exponents for spatial interaction is clearly consistent with the prior expectations of the gravity model and suggests that distance does have a significant impact on the volume of traffic between pairs of places. Furthermore, there is *some* evidence, comparable to Linnemann's 1966 study, that the distance exponent is greater in the more remote parts of Britain than in the more accessible and densely populated regions, with the implication that the remoter areas suffer a bigger adverse distance effect.

However, such interpretation of the distance exponents runs up against a number of problems, which centre around two issues. First, to what extent are the spatial variations in distance exponents an artefact of the model specification? Second, is the distance-decay effect to be interpreted as a manifestation of the price elasticity of demand, or the result of stochastic processes operating in the context of the given spatial pattern of origins and destinations? The problem of model specification arose following Curry's (1972) suggestion that distance parameters are affected by map pattern, and in particular the location of an origin (destination) on the edge of the map or somewhere more 'central'; in the British case, areas adjacent to the coast are by definition on the edge of the map, with transactions taking place in a limited number of directions. The ensuing discussion, concluded by Cliff *et al.* in 1976, showed that the apparent map effect was in reality the result of misspecifying the model to be fitted to the data. Gordon (1976) conducted some experiments for freight originating in twenty zones within Britain destined to fifty-three destination zones, these representing a collapsed version of the seventy-eight used by Chisholm and O'Sullivan. These experiments:

> suggest that Chisholm and O'Sullivan's results exaggerated the variation between origins in distance parameter values by a factor of about 2.5. Indeed, comparison of the range of values with the standard errors for individual regression estimates of the parameter leads one to doubt whether any of the deviations from the mean value are statistically significant.
>
> (Gordon 1976: 33)

The force of this comment may be illustrated as follows. Chisholm and O'Sullivan obtained distance parameters around the mean of –2.4, in the

Table 6.9 Great Britain: comparison of distance exponents and mean haul, originating road freight, 1964

	Sextile means values	
Sextile, distance exponent	*Distance exponent*	*Mean haul, kilometres*
1	–3.4	68
2	–2.9	60
3	–2.5	48
4	–2.1	48
5	–1.8	47
6	–1.5	43
Overall mean	–2.4	51

Source: Chisholm and O'Sullivan 1973: 132–3

range –1.3 and –4.8 (respectively Holborn/Charing Cross in London and Aberdeen). If that variation were exaggerated by a factor of 2.5 on account of model specification, then spatial differences in the distance coefficient would be small indeed, not least because the Aberdeen value is an extreme one; the next highest value is –3.8, for Bristol and Norwich. Of equal interest, though, is a comparison of the distance parameters with the mean length of haul for freight from each origin, this length of haul being estimated directly from the flow matrices (Table 6.9). The seventy-eight zones have been divided into six equal groups in descending value of the distance exponent. On the face of it, one might expect zones with a large value for the distance exponent to have relatively short mean hauls, since the distance exponent is generally treated as a measure of the distance-decay of transactions. In fact, the reverse situation obtains, a longer mean haul being associated with higher values for the distance exponent. The striking feature of Table 6.9, however, is the fact that there is very little variation in the mean haul between the third and sixth sextiles, and that the difference between them and the first two is quite small. Given that the overall standard deviation around the mean value of 51 kilometres is 14 kilometres, it is clear that Gordon's view that the spatial variation in distance exponents may not be statistically significant has considerable force. These doubts are reinforced when one notes that although the upper sextiles generally comprise peripheral zones and the lower sextile zones that are more centrally located, there are significant exceptions to this generalization. For example, the first two sextiles include, in addition to Bristol and Norwich, Bedford, Huntingdon, Leicester and Oxford, all of which occupy central locations in Britain. Conversely, the lowest two sextiles include peripheral zones such as Lancaster, Liverpool, Manchester and Newcastle-upon-Tyne. Clearly, notwithstanding the fact that some 48 per cent of the variance in distance exponents for originating traffic could be ascribed to the influence of generalized accessibility, measured by the logarithm of population-miles, there must be grave reservations about the significance of the spatial patterns obtained from these gravity-model outputs.

These reservations are reinforced by Gordon's finding that the data used by Chisholm and O'Sullivan showed a clear difference between originating flows and terminating flows:

> It can be seen that the inter-zonal variance in the distance parameter was very similar in scale for origins and destinations – and low in relation to the standard errors of the individual parameter estimates. However, inspection of the individual estimates and correlation with indices of economic potential both make it clear that the *systematic* variation between zones is predominantly in respect of origins. No spatial pattern was evident to the variation between destinations.
>
> (Gordon 1976: 32)

This finding led Gordon to conclude that 'we must look to differences in traffic composition rather than in the price sensitivity of consumers for a substantive explanation' of the distance-decay phenomenon. One way of addressing this problem is to consider the tonnage of freight generated per person resident in a zone of origin. A large tonnage implies either that the goods produced are bulky and of low unit value, and therefore price sensitive in respect of transport costs, or that there is a large volume of goods being distributed from warehouse/wholesale depots. Correlation of the origin distance exponents with tonnage per person and with generalized accessibility yielded almost identical levels of association. Given that the tonnage index is a much less appropriate tool for measuring commodity composition than is the accessibility index for measuring relative location, it is probable that this particular test of the commodity effect understates its importance (Gordon 1976).

This expectation was confirmed by examination of an entirely different data set for about 400 industrial establishments in the Severnside region of England, gathered in the latter half of the 1970s (Gordon 1978). Forty-four commodities were identified, for which the value–weight ratio accounted for between 24 and 40 per cent of the variance in distance exponents. Therefore, commodity characteristics are undoubtedly of considerable importance, not least because the number of supply points (origins) and consignment destinations may be limited. Perhaps the most important conclusion from this study, though, concerns the role of transport costs as a share of total distance costs: the indirect estimate is that transport costs account for no more than 10–15 per cent of the total. If that estimate is anywhere near accurate, we have a very interesting problem to solve.

Lin and Hanson (1976) defined transfer costs to include freight charges, in-transit inventory costs and the cost of protection against the unpredictability of the time in transit, and Gordon widened this concept to include information costs not associated with individual transactions. Lin and Hanson were working with United States data; they quote a range of figures for the share of total transfer costs accounted for by transport, indicating 20–25 per cent as the probable order of magnitude. Gordon argues that his figure of 10–15 per cent is consistent with the higher figure for America on account of the wider definition that he employed for transfer costs. On the other hand, we have already seen that within Britain the distance over which goods are shipped is, in general, very limited. As a consequence, inventory and uncertainty costs, and the cost of information, ought to be comparatively small, so it seems implausible to surmise that 85–90 per cent of transfer costs arise from the factors mentioned by Gordon. Therefore, there must be doubt about the strictly economic interpretation of distance exponents and the implied distance costs derived therefrom.

There clearly are enormous difficulties both in calibrating freight gravity models and in interpreting the resulting outputs in economic terms. It may well be the case, therefore, that one must call in question the assumption that distance exponents can be interpreted as representing the effects of consumer price elasticity of demand, and more willingly accept the proposition by Wilson (1967) that the basis for the gravity model lies in the probabilistic treatment of heterogeneous flows which involve diverse opportunities. If constraints are posited that the total flows from (to) an area account for all the traffic originating (terminating) in that area, and also for the total distance costs in the system, and if individual flows are then allocated between origins and destinations in a random manner, then it can be shown that a doubly constrained gravity model is the most probable distribution of flows. Such an interpretation does not deny the rationality of choice of either supplier or purchaser in arranging a particular shipment, within which transport costs will be one element affecting decisions: it does emphasize, however, the very large number of variables that is relevant in making those choices.

If we still insist on interpreting the distance exponents in strictly economic terms, we can conduct the following experiment. The Chisholm and O'Sullivan (1973) study provides estimates for each area of origin, derived directly from the origin–destination matrix in conjunction with a distance matrix, of the mean distance goods are shipped by road. Information is also provided on the structure of road freight charges (see also Chisholm 1971). For the sake of simplicity, let us assume that for each origin all the freight is shipped exactly the same distance, i.e., the mean haul. It is then a straightforward matter to calculate the difference in the estimated transport cost for origin zones, compared with the national average. For this purpose, the mean hauls for the sextiles in Table 6.9 have been used, supplemented by the extreme values at either end of the range. Note that although Glasgow Central and Sheffield had the lowest mean haul for originating traffic they do not have the lowest distance exponents.

Table 6.10 shows very clearly that the cost significance of the differences

Table 6.10 Great Britain: imputed differences in road freight costs for originating traffic (GB = 100)

Sextile, mean distance, kilometres	*Transport cost as % of GB mean*
68	109
60	104
48	98
48	98
47	97
43	96
Extreme values	
111 Aberdeen	132
32 Glasgow Central Sheffield	90

Source: Chisholm and O'Sullivan 1973: 120, 132–3

in mean haul, comparing one part of the country with another, is in fact very small. When it is further remembered that transport costs are but a small proportion of total costs, it seems likely that the interpretation of gravity-model distance exponents as a result of consumers' price elasticity of demand is implausible. Thus, it is evident that close examination of the results from the gravity modelling of freight flows within Britain, set in the wider context of gravity modelling more generally, serves to confirm and not to contradict the findings presented earlier in this chapter, to the effect that transport cost differences between the regions are not a major factor in determining relative prosperity. Furthermore, the interpretation of gravity models here presented at the intra-national scale is consistent with evidence at the international scale, discussed in Chapter 3.

CONCLUSION

If transport is very costly, as used to be the case, economic activity must be widely dispersed and situated where the natural resources are located. If the real cost of transport falls, it becomes increasingly possible to realize the available economies of scale in production, both internal economies and external economies. It is this situation which forms the kernel of the process leading to the growth of large agglomerations by cumulative causation. However, if access costs continue to fall in real terms, and were they ever to become zero, then there would be no need for activities to be close together, and agglomerations could break up and disperse. The limiting situation of zero transport costs could never be reached. However, it is possible that transport costs may become so low that the balance of advantage lies with dispersal rather than with spatial concentration, although the impact may vary with the nature of the activity and hence the role of the natural environment (Chisholm 1962, 1963). In the case of farming activities, low transport costs may lead to the localization of specialist producers where the physical environment is specially suitable. In contrast, low transfer costs make the dispersal of industry and financial services a real possibility.

The evidence which we have reviewed for Britain suggests very strongly that transport costs are now so low, and are so little affected by inter-regional location, that the location of a region is a minor factor affecting its relative prosperity. This conclusion is entirely consistent with the finding, reported in Chapter 3, that distance plays a very small part in shaping the global patterns of trade, with the consequence that Britain's location relative to Europe is not a material handicap.

However, while this conclusion is clearly relevant to the historical period with which we are concerned in this book, there are two further matters we should note. It is now becoming widely recognized that road transport creates serious adverse externality effects. Modern roads sterilize large swathes of land and have an undesirable impact on adjacent property and indeed upon the wider landscape, a *cause célèbre* being the gash through Twyford Down, near Winchester, to accommodate the M3. In addition, the emission of carbon dioxide and other products of combustion from the internal combustion engine is having a serious impact on the atmosphere, and the public mood in Britain and other developed countries is swinging towards the limitation of vehicle use to curtail acid rain, global warming, etc. Germany is leading the way with proposals to tax lorries using the motorways in that country (*Independent*, 10 February 1994). The effect will be to start raising the costs of road transport as these externality effects are internalized and charged to road users. A similar pressure is evident with respect to congestion, which manifestly is not solved by building ever more roads. To the extent that these pressures become more intense we may be entering an era in which more of the long-haul traffic is carried by rail, involving new investment for efficient modal interchange, or logistics facilities (Höltgen 1992).

Finally, this chapter has been concerned with measurable costs and has not directly addressed the question of time. When Oberon asked Puck to obtain 'a little western flower' he had an urgent, even if unworthy, need. It may be argued, therefore, that it is misguided, even irrelevant, to concentrate on costs. There are three responses to such a charge. First, from the point of view of an individual firm, the aspect of transport which can be compared with other expenditure and revenue is the cost incurred. Second, as we have seen, terminal costs account for a large proportion of total transport costs, and these terminal costs must be associated also with a significant proportion of total transit time. Third, for any consignment sent on a long journey, the critical thing is access to the trunk motorway or railway network. Once the consignment is moving along the trunk network, distance becomes a matter of relatively little moment. The crucial consideration, therefore, is not proximity but the reliability with which the transport operation functions, which is primarily dependent on the quality of the trunk network and the efficiency of terminal facilities. If the transport system is reliable, firms can organize their production and handling schedules to maximum advantage.

7

BRITAIN'S REGIONS, PORTS AND AIRPORTS

> In the conceptual metamorphosis of the transport system, from a unimodal basis to a multimodal approach, cargo movements came to be viewed and analysed in the light of the total distribution system.
>
> (Hayuth 1992: 203)

On the face of it, the dramatic reorientation of Britain's overseas trade towards Europe in the post-war period should favour the southern and eastern regions of Britain, as the relative advantage of their location has thereby been considerably improved. Such an expectation finds apparent confirmation in the replacement of Liverpool and London as Britain's two most important freight ports by Dover and Felixstowe, and is consistent with the fact that most of the Assisted Areas lie in northern and western Britain. It is easy to attribute the persistence, even if not the emergence, of the so-called North–South divide at least in part to the assumed advantage which the South has on account of proximity to Europe.

However, these inferences imply a number of causal links. In this chapter we will explore the available evidence to see whether these links do in fact behave in the manner postulated. The main part of the chapter will be concerned with freight but it is also relevant to consider some aspects of passenger traffic through the ports and airports, not least as the context for considering the potential impact of the Channel Tunnel (Chapter 8).

THE OVERSEAS TRADE OF BRITAIN'S REGIONS

In considering exports from Britain, it would be extremely useful if data were available on a regular basis which identified the region of origin for consignments, the ports through which those consignments were shipped and the country to which they are consigned. Equally, it would be helpful to have the equivalent data for imports. If such information were available, tabulations would be possible of the kind presented schematically in Figure 7.1. In this diagram, an export consignment from region A might be sent to one of three countries, and for each country there are three possible ports to

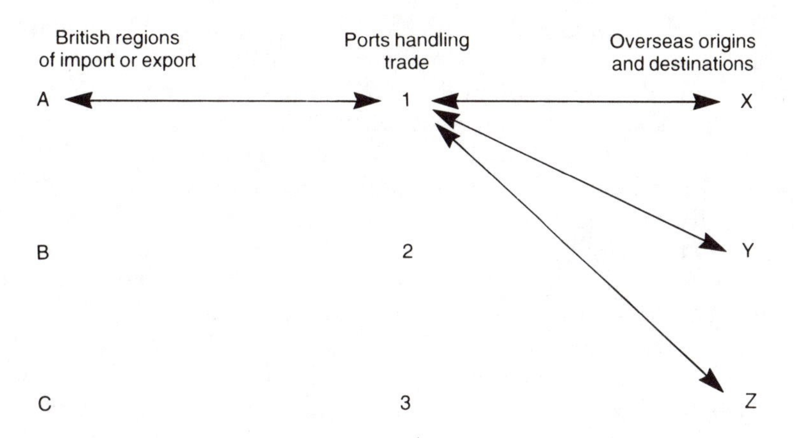

Figure 7.1 Schematic presentation of trade links

use. If proximity really is important, then we would expect the exports of regions to be biased towards countries that are nearby *and* simultaneously that the traffic with all overseas countries would use the port which is either nearest the British region of origin (destination) or nearest the overseas trade partner. Unfortunately, the requisite data for a full analysis of these propositions are not available and it is possible to examine only certain elements in the system sketched in Figure 7.1, and only for selected years (Chisholm 1985b, 1992; Hoare 1986, 1988, 1993).

Data are available for the three years of 1978, 1986 and 1991, for the amount of trade conducted by each of the regions of the United Kingdom with overseas countries or groups of countries. The information for the two earlier years is for tonnage only, whereas the 1991 data are for both tonnage and value. Chisholm's 1985 examination of the 1978 data identifies 'a somewhat perverse' pattern of trading relationships, in which it was the peripheral regions plus East Anglia that traded to a disproportionate extent with the nearby Continental countries, especially in the export of manufactures. Twenty-nine per cent of all non-fuel commodity exports by weight from the United Kingdom went to Belgium, France, Germany, Luxembourg and the Netherlands: the four regions with a share greater than this average were East Anglia, North, South West and Wales. If attention is focused on manufactures alone, then the four regions just identified, plus Scotland, are the regions which had an above-average propensity to trade with the 'near Continent'. The position with respect to imports was less clear.

The changes which occurred between 1978 and 1986 have been examined by both Chisholm (1992) and Hoare (1993), albeit using slightly different bases for the comparison. Both authors confirm that some elements of the 'perverse' situation in 1978 have been ameliorated or eliminated, most notably that the reliance of the South East on 'deep sea' traffic has diminished sharply. Table 7.1 reproduces the relevant regional tabulation for imports, and the export data are shown in Table 7.2. These and subsequent tables have been compiled on the following basis, which may be illustrated by reference to East Anglian imports from the EUR6 countries, shown in Table 7.1. The share that these imports are of total regional imports has been calculated, and that share has then been expressed as a percentage of the national proportion. Thus, at 190 in 1978, East Anglia's imports from the EUR6 countries were almost twice as important as for the United Kingdom as a whole. By 1986, the difference from the national position had diminished, as shown by the reduction in the percentage from 190 to 150. In this table, the regions have been ranked by the 1986 percentages, as also in Table 7.2 for exports. Although Northern Ireland is included in the United Kingdom total, the Province has been excluded from the regional tabulations because there are considerable difficulties with the data, notably with respect to land-based transactions with the Republic of Ireland.

In his analysis, Hoare interpreted the changes between 1978 and 1986 as reflecting the realignment of the southern part of the country towards the nearby Continental countries following the accession of Britain to the EU in 1973. However, he relied on an aggregation of the data for the South East and East Anglia, and for that grouping plus the South West, and used export traffic only. As Table 7.1. shows, the individual regions in those groupings have behaved rather differently, and there are also some important differences between the export and import data. More generally, Tables 7.1 and 7.2 display a complex pattern, both for the static patterns in the two years and for change in the intervening period. In general, the eight years to 1986 saw a closing of the regional differences, especially with respect to trade with the EUR6 countries. In other words, the regional trade patterns became more similar over time.

But that finding may depend either on the use of tonnage data, or the element of randomness associated with figures for single years. Unfortunately, the 1991 data are organized on a somewhat different basis from the information for the two earlier years, so that comparisons across the years must be approached with caution. However, if we compare Tables 7.1 and 7.2 with Table 7.3, it is immediately apparent that East Anglia remains particularly orientated to Europe for both imports and exports, and that the same is true of London within the South East region. Similarly, the North, and Yorkshire and Humberside, have remained at or below the national average of dependency on Europe. With the other regions, however, two points stand out. First, there is a marked contrast between the significance of Europe for imports as compared with exports, as with the North, North West and West Midlands in all three years. Second, for some regions there is considerable fluctuation in the relative importance of Europe for either imports or exports, e.g. Scotland and Wales. These instabilities suggest that

Table 7.1 United Kingdom: imports by tonnage, all non-fuel commodities, regional shares as percentage of UK average by area of origin (1978 upper figure, 1986 lower figure)

	EUR6	*Deep-sea*	*Scandinavia and Baltic*	*Iberia and Mediterranean*	*Eire*	*Central Europe*
East Anglia	190	38	153	103	57	92
	150	41	131	95	30	156
East Midlands	149	72	116	78	67	208
	144	54	125	61	40	219
West Midlands	198	44	123	68	67	333
	140	55	96	113	95	281
South East	128	84	97	138	70	150
	125	65	104	134	130	125
North West	106	101	87	110	110	100
	115	80	96	134	105	119
Scotland	76	114	126	45	27	42
	110	90	125	66	50	56
South West	114	79	86	268	93	58
	79	125	80	124	100	75
Yorkshire and Humberside	62	127	105	53	37	67
	70	125	133	53	80	63
Wales	34	156	73	52	13	33
	53	181	35	86	65	56
North	91	93	117	187	10	50
	44	166	64	113	155	90
United Kingdom	22.5	47.5	19.8	6.0	3.0	1.2
(% imports from)	33.8	36.8	17.8	8.0	2.0	1.6

Note: Northern Ireland has been excluded from the regional tabulations because the survey data exclude overland traffic with the Republic of Ireland
Source: Chisholm 1992: 566

Table 7.2 United Kingdom: exports by tonnage, all non-fuel commodities, regional shares as percentage of UK average by area of destination (1978 upper figure, 1986 lower figure)

	EUR6	*Deep-sea*	*Scandinavia and Baltic*	*Iberia and Mediterranean*	*Eire*	*Central Europe*
East Anglia	137	77	129	49	67	67
	122	36	126	125	82	62
South East	79	148	42	94	78	72
	119	83	65	117	109	48
South West	154	26	183	128	39	156
	113	34	193	75	97	138
East Midlands	82	115	118	74	100	111
	110	90	66	119	91	138
Scotland	107	102	103	109	56	83
	105	114	108	78	30	67
North	126	87	114	107	35	56
	102	121	109	62	48	129
Yorkshire and Humberside	93	112	88	119	68	133
	100	139	95	58	55	81
West Midlands	93	123	72	89	83	111
	85	138	66	97	121	176
Wales	118	85	137	64	70	117
	81	101	97	142	100	162
North West	72	108	109	105	154	106
	59	142	95	121	230	90
United Kingdom	32.1	36.3	13.5	8.1	8.2	1.8
(% exports to)	40.3	24.0	14.7	15.6	3.3	2.1

Note: Northern Ireland has been excluded from the regional tabulations because the survey data exclude overland traffic with the Republic of Ireland
Source: Chisholm 1992: 566

Table 7.3 United Kingdom, 1991: all non-fuel commodity imports and exports by tonnage, regional shares as percentages of the UK average by area of origin or destination

	Imports from			*Exports to*		
Region	*Rest of EU*	*Other short sea*	*Deep-sea*	*Rest of EU*	*Other short sea*	*Deep-sea*
East Anglia	143	103	45	115	94	55
West Midlands	139	99	52	87	87	152
South West	131	61	77	90	241	45
East Midlands	120	129	63	94	92	125
South East (excl. London)	114	97	83	94	81	130
North West	108	82	97	78	117	163
London	107	124	81	112	76	75
Scotland	94	101	108	107	117	67
Yorkshire and Humberside	75	158	107	94	97	123
North	64	118	138	99	64	126
Wales	48	31	193	114	72	72
UK average (%)	47.1	15.3	37.6	66.7	12.9	20.4

Source: Department of Transport 1993: 33

Table 7.4 United Kingdom, 1991: manufacturing imports and exports by tonnage, regional shares as percentages of the UK average by area of origin or destination

	Imports from			Exports to		
Region	*Rest of EU*	*Other short sea*	*Deep-sea*	*Rest of EU*	*Other short sea*	*Deep-sea*
East Anglia	118	76	64	94	126	104
West Midlands	113	85	73	87	95	139
South West	110	65	105	89	61	150
South East (excl. London)	102	96	97	106	96	86
North West	99	70	138	87	117	130
North	98	102	106	98	65	124
Wales	97	102	107	116	83	63
East Midlands	95	117	98	87	132	121
Yorkshire and Humberside	94	143	73	94	117	110
Scotland	93	96	127	90	224	67
London	83	122	134	116	64	73
UK average (%)	62.0	19.9	18.1	65.8	11.2	23.0

Source: Department of Transport 1993: 36

there are considerable variations in the trade flows on an annual basis. Comparison of the import and export patterns demonstrates pretty clearly that if a region has a high propensity to import from Europe it certainly does not follow that there will be a similar propensity for exports, and vice versa.

These tonnage data for 1991 can be further examined by separating out the traffic in manufactures, as in Table 7.4. The first striking feature of this table in comparison with Table 7.3 is the greater overall significance of the EUR12 countries for the import of manufactures than for non-fuel total imports; in contrast, there is little difference for exports, since British exports are in any case dominated by manufactured goods. The rank ordering of regions by relative import shares in Table 7.3 and 7.4 is similar, with the exceptions of London, Scotland and Wales, most notably London, which of all the regions is the least dependent on EUR12 for imports and the most reliant on deep-sea origins. With the exception of the East Midlands and London, the manufacturing import data for 1991 confirm the impression given by aggregate non-fuel imports, that southern and eastern Britain is more orientated to Europe than are the other regions. However, the manufacturing export data present a somewhat different picture. London and the South East both have a larger share of their exports destined to EUR12 than is the case for Britain as a whole, but the same is true of Wales. All the other regions have below-average trade shares with EUR12, including East Anglia, and the East Midlands and West Midlands share bottom place with the North West.

In sum, the manufacturing trade data do not display a clear-cut pattern if both import and export data are examined; the import data show the southern regions as being orientated to Europe, but this feature disappears with the export data. If we compare the trade in manufactures in 1991 with the position in 1978 (Chisholm 1985b: 974–5), then it does become apparent that the 'perverse' position that obtained in 1978 as shown by tonnage data, whereby the northern and western regions traded disproportionately with nearby Europe, has disappeared.

It is also possible to examine the value of trade for 1991 on the same basis as tonnage. Table 7.5, for all non-fuel trade on a value basis, may be compared with Table 7.3. The majority of the regions retain their relative position on the basis of imports from the rest of the EU, although there are three notable exceptions: Yorkshire and Humberside has moved from ninth position to third, while both London and the South West slipped down the ranking by four places. As for exports, the value data confirm the importance of European trade for East Anglia, London and Wales, but add to that list the North and the South West. There is also confirmation that whereas four regions have an above average proportion of imports derived from the EU, the significance of the Community for their exports is below average – East Midlands, North West, South East and West Midlands. Thus, although the value and volume data show a broad similarity of pattern, there are also substantial differences both within and between Tables 7.3 and 7.5, and quite

Table 7.5 United Kingdom, 1991: all non-fuel commodity imports and exports by value, regional shares as percentages of the UK average by area of origin or destination

	Imports from			*Exports to*		
Region	*Rest of EU*	*Other short haul*	*Deep-sea*	*Rest of EU*	*Other short haul*	*Deep-sea*
West Midlands	128	77	67	101	88	103
East Anglia	124	76	74	104	113	89
Yorkshire and Humberside	110	136	74	96	135	94
North West	105	84	98	85	101	126
South East (excl. London)	105	64	105	97	91	109
East Midlands	102	111	94	87	104	120
South West	100	62	112	112	84	85
North	95	81	113	111	95	82
Scotland	84	88	126	90	113	113
Wales	82	80	133	124	87	63
London	78	189	102	104	110	90
UK average (%)	51.6	12.1	36.3	56.0	11.2	32.8

Source: Department of Transport 1993: 33

Table 7.6 United Kingdom, 1991: manufacturing imports and exports by value, regional shares as percentages of the UK average by area of origin or destination

Region	Imports from			Exports to		
	Rest of EU	*Other short haul*	*Deep-sea*	*Rest of EU*	*Other short haul*	*Deep-sea*
West Midlands	131	77	65	101	88	103
East Anglia	122	73	78	99	108	98
Yorkshire and Humberside	115	118	73	94	139	96
North West	106	84	97	84	96	128
South East (excl. London)	105	62	106	97	90	109
North	101	70	109	114	104	75
South West	100	61	113	111	75	90
East Midlands	100	110	96	84	102	126
Wales	86	86	125	127	88	59
Scotland	83	82	129	90	128	107
London	76	202	98	104	109	89
UK average (%)	50.8	12.7	36.5	55.6	11.4	33.0

Source: Department of Transport 1993: 36

definitely the absence of a clear-cut spatial bias in the relative importance of trade with Europe.

Finally, Table 7.6 presents an analysis of the trade in manufactures on a value basis, which shows very little difference from the picture displayed in Table 7.5 for all non-fuel commodities by value. However, just as there are substantial differences between Table 7.3 and Table 7.5 for all non-fuel trade in volume and value terms, so also there are differences between Table 7.4 and Table 7.6 for manufactures only. These differences are very similar to those that have already been noted, so it would be otiose to repeat them. Perhaps the key feature to note about Table 7.6 is that the ranking of regions by the relative importance of the EU for imports differs sharply from that for exports. For example, whereas the two regions for which EU manufacturing imports are much more important than the average are the West Midlands and East Anglia, the two ranked highest by exports are Wales and the North. Taking Table 7.6 as a whole, there is little evidence to suggest that the regions which are nearest to Europe systematically trade with our neighbours on a scale greater than the national average.

Taking the information for 1978, 1986 and 1991 together, East Anglia and the West Midlands are consistent in having a high proportion of imports drawn from Europe but they are much less consistent with respect to exports. In contrast, Scotland and Wales generally appear in the lower part of the tables on the import data and, in the case of Scotland, on the export data as well; in contrast, Wales has strong export links with Europe. The other regions display considerable variation in their rank-order positions and also in the behaviour of their imports and exports. And finally, the 1991 figures show London performing very differently from the rest of the South East and also East Anglia. There is nothing in the data which we have examined to suggest that there is a clear-cut advantage for the regions of southern and eastern Britain on account of proximity to Europe, as manifested in their trade patterns.

PORT HINTERLANDS

A limited amount of information is available for the proportion of regional imports and exports routed through ports located in the regions. In his examination of 1964 and 1978 exports, Hoare was able to allocate individual ports to the standard regions of freight origin and destination. The data available to Chisholm required a somewhat cruder analysis, since the assignment of flows was to groups of ports, necessitating some amalgamation of regions, to give seven in total. Thus, the figures of 48 and 58 per cent for 1978 in Table 7.7 have been derived on different bases. There seems little doubt that between 1964 and 1978 there was a sharp reduction in the proportion by tonnage of exports shipped through a 'local' port, but that there was some reversal of that trend between 1978 and 1986 for all non-fuel

Table 7.7 United Kingdom: percentage of regional imports and exports using ports in the same region or amalgamation of regions

	Tonnage			Value
	1964	*1978*	*1986*	*1986*
Exports[1]	63	48	–	–
Exports[2] all non-fuel commodities	–	58	61	32
Imports[2] all non-fuel commodities	–	69	65	36
Exports[2] non-fuel general cargo	–	34	28	–
Imports[2] non-fuel general cargo	–	43	36	–

Sources: [1] Hoare 1986: 35
[2] Chisholm 1992: 564

commodities. However, there was some decline in the significance of 'local' ports in respect of imports. If the analysis is restricted to non-fuel general cargo, which approximates to manufactures, then the downward trend in the share of trade using 'local' ports clearly continued between 1978 and 1986 for both imports and exports, and stood at about one-third of the tonnage in 1986. However, in value terms, the significance of 'local' ports is much less than is the case when tonnage data are used. So far as manufactured imports and exports are concerned, it is probable that well under one-quarter by value of shipments passes through a 'local' port.

The structure of inland freight charges contributes to this situation. As noted in Chapter 6, container traffic is charged on a basing-point system – as if it were routed through the nearest port with an appropriate shipping service, even though it may be consigned through another, more distant port. In the early days of unitized traffic, containers accounted for over 60 per cent of the total tonnage of unitized freight but this proportion fell until about 1980. Since then, containers have maintained their share, at about 40 per cent. (*Port Statistics* 1991: 44). Thus, within the unitized sector a substantial proportion of traffic is handled under basing-point rules, a fact which must contribute to the rather loose links which ports have with their immediate inland areas. That this is indeed the case is attested by the anxiety felt in Scotland, that container traffic which originates or terminates there does not use Scottish ports as much as might be expected, being routed in fact through ports south of the Border.

The evidence is quite conclusive, that ports are no longer dependent on their immediate hinterlands and that they compete in a market which is 'far more fluid and competitive than previously. This has obvious implications for ports to adopt go-getting marketing strategies, with penalties in a loss of trade for those failing in the chase' (Hoare 1988: 1358). That being the case, the spotlight shifts to the role that may be played by the location of a port with respect to overseas trading partners (the foreland of a port); or to the characteristics of the port itself – its efficiency, the frequency of sailings, etc. – and it is to these questions we now turn.

PORT FORELANDS

The realignment of Britain's trade towards Europe, and especially EUR6, suggests that ports within Britain which are near the mainland should be at an advantage relative to those which are further away. Consequently, one would suppose that the south-eastern ports should have gained market share in the post-war period. At the aggregate level, there has been no such shift (Table 7.8). However, the aggregate comprises two contrary trends. The relative importance of the Southampton–Wash ports for bulk fuels (liquids) has declined quite sharply, though in recent years the share has stabilized at 17–20 per cent. In contrast, these ports have gained in importance for the 'other traffic', which mainly comprises dry-bulk and semi-bulk goods and unitized traffic, though the gain appears to have petered out recently. The South East's loss of market share in the case of fuel is closely linked with the exploitation of the North Sea for petroleum, leading to a large expansion in trade for Scottish ports in particular. Table 7.8 does suggest, therefore, that the south-eastern ports have indeed benefited from the post-war changes in the geography of Britain's trade, if bulk fuel is discounted.

Table 7.9 shows that bulk and semi-bulk traffic has accounted for a very large proportion of Britain's overseas traffic – nearly 90 per cent in 1965 and still over three-quarters in 1992. Conventional traffic – manhandled and winched on and off the vessels – has dwindled from just over 10 per cent of

Table 7.8 The changing relative importance of Great Britain's ports from Southampton to the Wash: imports plus exports by tonnage

	1976	*1980*	*1985*	*1989*	*1990*	*1991*	*1992*
Total tonnage, Great Britain (million tonnes)							
Bulk fuel[1]	213	268	280	131	144	149	155
Other	122	144	169	168	173	172	173
Total	335	412	449	299	317	321	328
Ports: Southampton to Wash (% tonnage)							
Bulk fuel[1]	34.1	24.3	20.9	16.8	19.9	20.7	17.0
Other	33.2	42.6	43.2	45.9	46.0	45.3	46.0
Total	33.8	30.8	29.3	33.1	34.1	33.9	32.3

Note: [1] Just 'fuel' in 1976, and 'liquid bulk' in 1990, 1991 and 1992
Source: *Port Statistics*, various issues

Table 7.9 Composition of Britain's overseas trade through ports, by mode of appearance: percentage distribution by tonnage, imports plus exports

	Great Britain					*United Kingdom*			
	1965	*1970*	*1975*	*1980*	*1988*	*1988*	*1990*	*1991*	*1992*
Bulk fuels	52.6	59.3	56.8	57.9	50.1	49.7	49.0	50.4	50.4
Other bulk traffic	25.5	21.4	20.8	17.7	20.6	21.0	20.7	19.9	19.5
Semi-bulk traffic	9.4	7.7	6.7	6.5	7.6	7.7	6.9	6.5	6.1
Unitized traffic	1.5	5.6	11.1	15.0	20.9	20.8	22.5	22.5	23.3
Conventional traffic	11.0	6.0	4.6	2.9	0.8	0.8	0.9	0.7	0.7
All foreign traffic	100.0	100.0	100.0	100.0	100.0	100.0	100.0	100.0	100.0
Million tonnes	189	244	226	248	308	312	320	325	333

Source: *Port Statistics* 1992: 14

Table 7.10 Great Britain: percentage distribution of unitized traffic, imports plus exports, by tonnage

	1970	*1975*	*1980*	*1985*	*1989*	*1990*	*1991*	*1992*
All Thames, Kent Sussex and Hampshire	28.7	35.2	41.6	39.4	41.7	39.6	42.0	42.2
London	11.7	10.0	9.4	8.3	7.1	8.3	8.6	8.8
Medway	0.2	0.8	1.5	3.2	3.7	3.7	3.2	3.6
Ramsgate	0.3	0.1	0.1	2.1	3.4	2.7	4.4	4.8
Dover	9.3	12.5	16.4	17.2	18.6	15.3	16.1	16.4
Folkestone	0.0	1.4	1.2	0.9	0.9	0.8	0.7	0.2
Newhaven	1.1	1.4	1.1	1.4	1.4	1.5	0.7	0.2
Portsmouth	0.0	0.2	1.0	2.4	2.7	2.8	3.1	3.0
Southampton	5.6	8.8	10.3	3.8	3.8	4.5	5.3	5.1
All Wash, East Anglia and Haven	29.2	26.3	25.9	31.8	33.8	33.8	32.6	32.5
Great Yarmouth	1.3	2.1	2.0	1.9	1.8	1.7	1.1	–
Felixstowe	10.2	12.8	13.6	17.9	21.6	22.4	21.2	22.7
Ipswich	1.4	2.2	3.5	4.6	4.5	4.9	3.9	3.7
Harwich	16.1	8.9	6.6	7.0	5.5	4.3	5.7	5.6
All other	42.1	38.5	32.5	28.8	24.5	26.6	25.4	25.3
All GB	100.0	100.0	100.0	100.0	100.0	100.0	100.0	100.0
Million tonnes unitized freight	13.6	25.0	37.0	51.0	70.3	71.6	72.6	77.1

Source: *Port Statistics*, various issues

the total to a negligible level. In contrast, unitized traffic has not only replaced conventional handling, but has expanded to occupy almost one-quarter of the freight market. It is this sector, mainly comprising containers and roll-on/roll-off road vehicles, which warrants special scrutiny – not least because the unitized traffic is almost exclusively manufactures, and is, therefore, traffic which should be particularly sensitive to the changing relationship between Britain and Europe.

In 1970, the ports from Southampton to the Wash accounted for 58 per cent of unitized imports and exports by tonnage, and this share has risen to about 75 per cent in recent years (Table 7.10 and Figure 7.2). Most of the gain in market share has occurred along the coast from the Thames to Hampshire, most notably at Dover. In contrast, the East Anglian and Haven ports have made only modest gains, the huge advance of Felixstowe having occurred largely at the expense of Harwich.

On the face of it, the evidence in Table 7.10 does suggest very clearly that proximity to Europe has been a considerable advantage for ports in the post-

Figure 7.2 Great Britain: standard regions and port groups

war period. However, caution is necessary in drawing that conclusion. The 58 per cent of unitized traffic handled by the south-eastern ports in 1970 was a much larger proportion of that traffic than the proportion of Britain's trade conducted with EUR6 – almost three times in fact (Table 2.7 and Figures 2.5 and 2.6). In other words, these ports had a much larger share of the unitized traffic than would be expected from the then significance of Europe in Britain's trade. In contrast, whereas the south-eastern ports have since increased their share of unitized traffic from 58 to 75 per cent, this change has been much less than the doubling in the share of Britain's trade with the EUR6 countries that has occurred over the same period. The high concentration of unitized traffic on the South East in the early post-war period, and the slower-than-might-have-been-expected growth in their share (remembering that 100 per cent is of course a limiting state) does suggest that some further considerations must be explored.

If proximity to Europe does confer the advantages hypothesized, we would expect the ports in the South East to have a clear bias in their traffic patterns towards nearby European countries. Chisholm (1992) explored this possibility, using data averaged over the years 1985 to 1988. That analysis has been updated for the year 1992 (Tables 7.11 and 7.12). Unfortunately, the data on overseas origins and destinations for individual ports are for the aggregate

Table 7.11 Great Britain, 1992: overseas origins of total imports, percentage distribution, selected ports

	By value				*By tonnage*			
Port	*EUR6*	*Other short sea*	*Deep-sea*	*Total*	*EUR6*	*Other short sea*	*Deep-sea*	*Total*
Ipswich	89.2	5.4	5.4	100.0	80.9	8.3	10.8	100.0
Dover	82.6	16.9	0.5	100.0	80.6	18.1	1.3	100.0
Harwich	74.1	24.8	1.1	100.0	74.2	24.8	1.0	100.0
Hull	73.2	22.8	4.0	100.0	59.7	27.5	12.8	100.0
Forth	51.9	25.1	23.0	100.0	35.0	45.1	19.9	100.0
Grimsby and Immingham	44.9	46.8	8.3	100.0	17.5	43.7	38.8	100.0
London	44.6	17.2	38.2	100.0	36.3	28.1	35.6	100.0
Medway	44.5	15.5	40.0	100.0	52.9	26.7	20.4	100.0
Manchester	41.3	34.3	24.4	100.0	24.2	43.5	32.3	100.0
Tees	40.0	31.5	28.5	100.0	15.3	15.3	69.4	100.0
Felixstowe	29.9	13.4	56.7	100.0	28.8	20.2	51.0	100.0
Southampton	7.8	10.0	82.2	100.0	11.7	41.6	46.7	100.0
Liverpool	7.7	42.4	49.9	100.0	6.0	25.3	68.7	100.0
Great Britain	50.9	22.8	26.3	100.0	27.7	33.9	38.4	100.0

Source: *Port Statistics* 1992, Tables 3.10 and 3.11

Table 7.12 Great Britain, 1992: overseas destinations of total exports, percentage distribution, selected ports

	By value				*By tonnage*			
Port	*EUR6*	*Other short sea*	*Deep-sea*	*Total*	*EUR6*	*Other short sea*	*Deep-sea*	*Total*
Ipswich	79.3	10.2	10.5	100.0	59.4	30.6	10.0	100.0
Hull	76.8	16.0	7.2	100.0	62.4	24.5	13.1	100.0
Dover	75.8	19.0	5.2	100.0	80.7	18.2	1.1	100.0
Harwich	68.9	25.6	5.5	100.0	72.8	20.3	6.9	100.0
Forth	62.9	14.5	22.6	100.0	77.4	9.6	13.0	100.0
Tees	60.5	24.8	14.7	100.0	52.2	21.4	26.4	100.0
Medway	58.5	12.2	29.3	100.0	82.0	6.6	11.4	100.0
London	46.7	11.9	41.4	100.0	52.0	19.4	28.6	100.0
Manchester	40.1	25.8	34.1	100.0	48.3	43.7	8.0	100.0
Grimsby and Immingham	40.0	51.2	8.8	100.0	63.5	25.5	11.0	100.0
Felixstowe	25.5	15.6	58.9	100.0	26.3	19.7	54.0	100.0
Southampton	19.6	17.7	62.7	100.0	47.3	33.2	19.5	100.0
Liverpool	2.0	46.6	51.4	100.0	4.5	51.3	44.2	100.0
Great Britain	50.9	23.9	25.2	100.0	51.1	25.5	23.4	100.0

Source: *Port Statistics* 1992, Tables 3.10 and 3.11

of all freight, not just unitized traffic. Nevertheless, some clear conclusions can be drawn. Both tables have been arranged in descending order of the importance of the EUR6 countries for the trade of a port by value. The thirteen ports in these tables, the same as the ones previously analysed for 1985–8, accounted in 1992 for over 80 per cent of imports by value and 70 per cent by tonnage; the comparable figures for exports are over 80 and 50 per cent respectively.

Five ports stand out as having an above average share of both imports and exports in trade with the EUR6 countries – Dover, Forth, Harwich, Hull and Ipswich. Conversely, Liverpool trades hardly at all with these countries. But Felixstowe and Southampton, both ports which one might expect would trade heavily with nearby Europe, do not do so and in fact are orientated to deep-sea traffic. The position of Felixstowe, with a low dependence on EUR6, is striking, given its great success since the Second World War and its almost sole reliance on unitized traffic. The data for 1992 do not unequivocally confirm the expectation that there should be a clear geographical pattern to the overseas trade of ports according to location within Britain, especially when one compares Liverpool and Manchester on the one hand, and Felixstowe and Harwich on the other.

The changes which have occurred between 1985–8 and 1992 confirm that conclusion. Broadly speaking, the pattern has remained the same, but there have been some shifts in ranking to note. Medway has slipped from its position among the first five to the middle of the table, and Felixstowe, which formerly occupied a middle position, has moved to being third from the bottom. At the same time, London has moved sharply up the ranking to a middle position, while Forth has vacated that position to join the top five.

In sum, the evidence concerning the overseas trade of individual ports shows that there are huge variations in the geographical patterns and that these variations are not fully consistent with the inference that one may draw from the shift in Britain's trade towards Europe and the implied change in the relative advantages of ports located in the south and east compared with the north and west. Although relative location is one factor, it is clearly not the dominant consideration. As a result, we must consider what else may be important and it is, therefore, appropriate to consider whether issues relating to port efficiency may play a part.

PORT EFFICIENCY

It is generally accepted that Britain's ports had, by 1979, experienced a long period of overmanning and under capitalization and, as a consequence, high unit costs (Bassett 1993). It is also widely agreed that it was appropriate for the Conservative government elected in that year to institute reform with the aim of raising efficiency. Two strategies were adopted, privatization and deregulation. Many of Britain's docks used to be owned and managed by the British Transport Docks Board. In 1981, legislation provided for the privatization of these nineteen ports, which was duly effected in 1983 with their sale to a holding company, Associated British Ports. The second phase of privatization began with 1991 legislation which permits, but does not compel, the privatization of the trust ports. There were more than 100 of these ports, governed by individual and often archaic constitutions which placed them neither in the public nor the private domain. The effect of the 1981 and 1991 legislative changes is that most of the major ports should be in private ownership, replacing the mixed pattern of private, public and trust ownership which previously existed. However, progress in privatizing the trust ports has been slow (*Independent*, 21 December 1993).

The other major strand of policy in respect of the ports industry was the abolition of the National Dock Labour Scheme (NDLS) in 1989. This Scheme was introduced in 1947 in an effort to give dock-workers secure employment and decent work conditions, both of which were impossible with the pre-existing system of daily hire – or casual labour. Although the intentions which lay behind the introduction of the NDLS were entirely laudable, the Scheme had the unintended effect of creating highly protected employment which fostered job demarcation and inflexibility in the deployment of the workforce. This inflexibility was, in turn, a major deterrent to investment and innovation. However, the Scheme did not apply to all ports. At the time proposals were published in 1989 to abolish it, the Scheme applied to forty ports but thirty-five others were listed as being exempt (Secretary of State for Employment 1989: Appendix 1). The reasons for this distinction lie in the history of the dock industry, and in particular in the fact that when the Scheme was introduced in 1947 it applied to the docks which at that time were major handlers of freight. The packet ports of Dover and Harwich were exempt, as also Felixstowe, which was little more than a staithe on an expanse of coastal mudflats. Dover and Felixstowe were able to seize the opportunity thus presented. Of the thirteen ports listed in Tables 7.11 and 7.12, only these two ports plus Harwich were outside the NDLS scheme at the time of its abolition in 1989.

Although privatization was an important change in the structure of the port industry, it was the abolition of the NDLS which really unshackled the industry and put all the ports on a level playing field. Already by 1992, evidence was beginning to accumulate that the industry had become much more efficient and competitive, with the result that Dover and Felixstowe were under pressure in maintaining their market shares (Chisholm 1992). That assessment has been confirmed in the case of Bristol (Basset 1993) and the results are now showing up in changes in the rates at which productivity is improving in the various ports.

Annual data are available since 1983 for the number of workers employed in the groups of docks (see Figure 7.2 for group configurations), from which it is a straightforward matter to calculate the tonnes of freight handled per worker each year. This is a very crude output measure, because it ignores the commodity mix handled by each port: some specialize in bulk commodities and achieve large annual throughputs relative to their manpower; others are devoted primarily to unitized freight, where the throughput is lower. In addition, ports such as Dover and Harwich handle considerable numbers of passengers. Static comparisons between ports on the basis of tonnage handled per employee do not have much meaning. On the other hand, if the commodity structure remains fairly constant for individual ports or groups of ports, then valid comparisons may be made for the rate at which productivity changes over time. Tables 7.8 and 7.10 suggest that since the mid-1980s there have not been big changes in commodity mix for the ports between Southampton and the Wash taken as a group, so it is worth exploring what can be learned from changes in the tonnage handled per worker.

For this purpose, 1989 has been taken as the base year, equal to 100, this being the year that saw the end of the NDLS. As Table 7.13 shows, the south-eastern ports saw their throughput per employee rise by 42.8 percentage points between 1983 and 1988, whereas the increase elsewhere in Great Britain was 28.7 percentage points. In the run-up to the abolition of the NDLS, the ports outside the south-east were improving their productivity

Table 7.13 Great Britain: productivity changes in the ports

	Annual tonnage of freight handled per employee (1989 = 100)		*Unitized plus conventional traffic as percentage of total traffic by weight*[1]	
	Southampton to Wash	*Rest of Great Britain*	*Southampton to Wash*	*Rest of Great Britain*
1983	64.1	81.1	29.3	7.6
1984	76.2	89.7	28.5	7.9
1985	83.4	97.4	29.7	8.2
1986	89.1	104.8	30.1	8.0
1987	94.7	104.1	32.6	8.4
1988	106.9	109.8	32.4	9.0
1989	100.0	100.0	35.4	9.5
1990	111.7	117.5	35.1	10.2
1991	119.8	134.5	35.7	9.9
1992	127.1	146.5	39.4	10.4

Note: [1] Includes traffic with UK offshore installations, non-oil
Source: *Port Statistics*

less rapidly than the south-eastern ports. Since 1989, the position has been dramatically reversed, and it is the south-eastern ports which have been lagging.

Table 7.13 also contains summary data for the structure of trade. The port statistics distinguish the five categories of traffic shown in Table 7.9. Of these, the main distinction relevant in this context is between bulk and semi-bulk freight on the one hand, and unitized plus conventional freight on the other. The bulk commodities require less manpower than unitized and conventional traffic, so that any major shift in the balance between these commodity groups would imply shifts in recorded tonnage per employee independent of real changes in productivity. Table 7.13 shows that for both the south-eastern ports and the rest of the country there has been a progressive shift in the share of unitized plus conventional traffic in the total freight handled, and no trend change is evident within the period shown. Hence we may conclude that the dramatic change in productivity trends noted above reflects real changes in the efficiency of handling goods and is not an artefact of changes in the balance between high and low productivity types of traffic.

If we now refer back to Tables 7.8 and 7.10, it will be noted that the growth in relative importance of the south-eastern ports, which had seemed inexorable, has, since 1989, come to a halt. This is true for all freight other than bulk fuels (Table 7.8) and also for unitized traffic (Table 7.10). Furthermore, with respect to unitized freight, both Dover and Felixstowe have been struggling to maintain the market share they had won, a struggle of considerable importance for both ports since virtually all the freight they handle is unitized. We would not expect the privatization and deregulation of the docks industry to have an instantaneous impact on the relative fortunes of the ports, but the evidence is persuasive that Dover and Felixstowe both owed a substantial part of their success to the fact that they were not subject to the NDLS. With the demise of that Scheme, other ports elsewhere in the country that have for long been handicapped are now taking the fight to the nation's two leading ports for trade in manufactures. There is no certainty that they will be able to hold on to the market share they have won, even taking no account of the probable impact of the Channel Tunnel on Dover in particular.

CONCLUSION ON SEA FREIGHT

It seems quite clear that the location of a port is only one factor in its success, and probably a diminishing factor in many cases. We have seen that to an increasing extent ports compete nationally for trade in manufactures, and that in this sector of commerce, at least, the traditional concept of a port serving a geographically bounded hinterland is now of limited value. In the same way, it is clear that the overseas connections of a port, and hence the pattern of its trade, is only in part determined by its location. Thus, although Dover trades almost exclusively with the nearby mainland of Europe, Felixstowe is orientated primarily to the deep-sea trades, as is Southampton and, until recently, London. Thus, serious attention must be given to the ability of dock management to innovate and invest, to gain operational efficiencies. There is little doubt that the accidents of ownership have played a part, Felixstowe being a private port and Dover a trust port, compared with the public ownership of other major ports until recently. At least as important, though, has been the impact of the National Dock Labour Scheme; the abolition of this Scheme in 1989 has opened up a new era of competition, in which ports that had lost out show signs of regaining some trade share.

AIR FREIGHT

Thus far, no mention has been made of air freight. Despite the near doubling of the tonnage handled between 1982 and 1992, the total was only 1.2 million tonnes in the latter year for the whole of the United Kingdom. In the same year, the sea ports handled a total of 333 million tonnes of international freight, of which 78 million tonnes was unitized. In tonnage terms, air freight is a small component of Britain's international trade, though in value considerably more important than the tonnage figures indicate.

There are three features of the air freight traffic worthy of note in the present context. First, over 80 per cent of air freight is handled by the six airports which are located in and around London – City, Gatwick,

Heathrow, Luton, Southend and Stansted. Second, the freight is characterized by its perishable nature or by the high unit value of consignments and the premium that exists for speed. Third, a considerable proportion of the traffic is with countries outside Europe. As a result, it is not so much location within Britain that matters as the availability of fast and reliable overland transport links within Britain. Similarly, it is unlikely that the opening of the Channel Tunnel will divert much, if any, freight from the air to through rail freight services or Le Shuttle. Altogether, it seems unlikely that patterns of air freight have been much influenced by locational matters of a kind germane to the wider issue of Britain's position on the edge of Europe.

PASSENGER TRAFFIC

Table 7.14 shows the enormous increase that there has been in overseas passenger traffic since the mid-1950s – a more than tenfold expansion. This table also shows the dramatic shift that has occurred in the relative importance of air and sea traffic; passenger movements by sea used to

Table 7.14 United Kingdom: percentage distribution of passenger movements, arrivals plus departures

	1956	*1960*	*1970*	*1980*	*1990*	*1991*	*1992*
Ireland							
Sea	16.4	11.4	5.1	3.9	2.6	3.0	2.7
Air	4.6	5.7	5.0	2.8	4.2	3.8	3.8
Total	21.0	17.1	10.1	6.7	6.8	6.8	6.5
Europe plus Mediterranean							
Sea	40.5	37.4	28.7	31.9	25.3	27.3	26.0
Air	25.2	32.6	45.4	42.6	48.6	45.6	46.5
Total	65.7	70.0	74.1	74.5	73.9	72.9	72.5
Rest of the World							
Sea	7.7	4.6	0.9	–	–	–	–
Air	5.6	8.3	14.9	18.8	19.3	20.3	21.0
Total	13.3	12.9	15.8	18.8	19.3	20.3	21.0
World							
Sea	64.6	53.5	34.7	35.8	28.0	30.3	28.8
Air	35.4	46.5	65.3	64.2	72.0	69.7	71.2
Total	100.0	100.0	100.0	100.0	100.0	100.0	100.0
Total movements (million)							
Sea	5.9	6.9	11.5	23.4	29.7	31.0	33.0
Air	3.2	5.9	21.7	42.1	76.4	71.3	81.5
Total	9.1	12.8	33.2	65.5	106.1	102.3	114.5

Source: *Port Statistics*. Air traffic data for 1991 and 1992 from *UK Airports*, CAP 614; classification differs slightly from earlier years

account for almost two-thirds of all movements, but now account for less than one-third. This modal shift has been most extreme for destinations outside Ireland, Europe and the Mediterranean, with the complete cessation of passenger liner traffic. Nevertheless, seaborne passenger traffic has also lost share in both of the other markets identified in Table 7.14, though less dramatically than for the 'rest of the world'. Despite the loss of market share, the absolute volume of seaborne traffic has increased substantially, from 5.9 million movements in 1956 to 33 million in 1992. Meantime, air traffic expanded far more quickly, from 3.2 million to 81.5 million movements.

These summary data show quite clearly that in any discussion of overseas passenger traffic it is essential to discuss air and sea travel separately. Such a separation is also essential if one is to make any assessment of the impact that the Channel Tunnel may have on traffic flows and hence upon competing air and sea services. As we will discover in the following pages, the spatial characteristics of air and sea passenger traffic differ quite substantially, as also do the trends in the geography of movement. In a nutshell, sea passenger traffic has become progressively concentrated on south-east England and especially on services to France, whereas there has been no such shift in respect of air services, and if anything the reverse. However, these shifts are in the context that, already in the early post-war period, Europe dominated Britain's passenger traffic in a manner which contrasts sharply with freight traffic. Although mainland Europe has taken a somewhat increased share of total traffic since then, that increase has been at the expense of Ireland; meantime, again in contrast to the behaviour of the freight sector, the more distant parts of the world have been taking an ever-larger proportion of total passenger traffic – up from 13 per cent in the 1950s to 21 per cent in 1992.

PASSENGER TRAFFIC THROUGH THE SEAPORTS

Over the last quarter of a century, there has been a marked decline in the relative importance of the west coast ports (Table 7.15), reflecting the diminished relative significance of Ireland and the cessation of liner traffic across the Atlantic. The east coast ports, after a remarkable gain in market share in the 1960s, have also rather steadily lost share until recently. Up until the mid-1980s, the ports of the Thames and Kent sector of the coast were gaining in traffic more rapidly than other ports but more recently the growth in traffic has been fastest along the south coast. Perhaps the single most striking fact shown in Table 7.15, however, is the dominance of Dover. This port has consistently handled more than 40 per cent of all passenger traffic, and in recent years in excess of 50 per cent.

Dover is of special interest, not just because it is far and away the most important packet port in the United Kingdom but also because the ferry services through this port are most directly affected by the Channel Tunnel. Table 7.15 shows that Dover's share of all traffic was fairly stable until the

Table 7.15 United Kingdom: international passenger movements by sea, arrivals and departures, percentage shares

	1965	*1970*	*1975*	*1980*	*1985*	*1990*	*1991*	*1992*
Dover	42.5	44.6	41.9	47.0	52.6	52.5	51.7	54.5
All Thames and Kent	59.5	54.0	59.3	61.4	64.2	64.3	64.3	64.0
Newhaven	4.5	4.4	3.7	3.4	3.3	2.8	–	–
All South Coast	12.5	13.3	12.7	13.6	12.4	16.3	16.4	15.4
West Coast	17.5	13.9	11.2	10.7	11.0	9.3	9.8	9.5
Hull	0.3	1.5	1.8	2.1	1.7	3.3	–	–
Felixstowe/Ipswich	–	–	1.9	4.2	2.9	1.5	1.8	2.0
Harwich	7.5	10.9	9.2	7.2	7.4	4.5	4.0	5.5
All East Coast	10.5	18.8	16.8	14.3	12.4	10.1	9.5	11.1
All port areas	100.0	100.0	100.0	100.0	100.0	100.0	100.0	100.0
Passengers (million)	9.1	11.3	16.2	23.3	26.1	29.7	31.0	32.9

Source: *Port Statistics*

mid-1970s, that it gained about ten percentage points between then and the mid-1980s and has held its position since. Some of Dover's gain has been at the expense of the east coast ports, but a roughly equal part is attributable to the pressure Dover has exerted on the traffic of other ports in its vicinity, such as Folkestone; it was in September 1991 that Stena Sealink announced the cessation of its services from this port. Over the period shown in Table 7.15 there has clearly been some concentration of passenger traffic on the Thames and Kent ports and, within that sector of the coast, differentially upon Dover. In recent years, though, both the south coast ports, serving France and Spain, and the east coast ports have demonstrated their ability to compete with the shortest ferry route: Dover–Calais.

A very large part of the ferry traffic is of passengers with an accompanied car, mainly but not exclusively holiday traffic. For car-borne passengers, the progressive improvement of the road network in Britain and on the mainland of Europe has no doubt played some part in the concentration of traffic on the extreme south-east, since these improvements make it worth while to take a longer detour to Dover in order to gain from the short crossing time of the ferries plying from that port. However, the fact that the south and east coast ports are showing their competitive mettle indicates that the improvement in road communications is certainly not the only consideration. Probably equally important is the fact that Dover has a much greater number of sailings each day than is the case with other ports.

Overall two points are of main interest. After a period of rapid adjustment and change in the post-war period, in which there was some modest concentration of traffic on the extreme south-east, the broad geographical pattern has remained fairly stable in recent years. Despite the longer sea crossings, about a quarter of all passengers use east and south coast crossings, no doubt because the longer time at sea is offset by less driving in Britain and/or in mainland Europe. The second key point is the dominance of Dover, with more than half the total traffic, a position now under threat from the Channel Tunnel.

PASSENGER TRAFFIC BY AIR

International passenger movements handled by the airlines have been increasing at a far more rapid pace than journeys undertaken by sea; the airborne sector of the market is much the most buoyant. But this market is in fact two quite distinct markets, served by charter services and scheduled services. For all practical purposes, charter flights are exclusively undertaken for holiday purposes; therefore, the geography of this traffic is dominated by the main holiday destinations, mainly located in the Mediterranean and especially in Greece, Italy and Spain. In contrast, scheduled services cater for a mixture of business and holiday traffic, and the pattern of these services is strongly biased towards the main commercial and population centres of the world and of Europe. Thus, as a rough approximation we may equate the scheduled services with business travel, in contrast to the holiday traffic of charter services.

Table 7.14 shows quite clearly that, in aggregate, the most rapid expansion of air travel has been with countries further afield than Europe and the Mediterranean. In the early post-war years, this relatively rapid growth in traffic with more distant countries was partly the substitution of air travel for the ocean-going liners, but that substitution effect had been exhausted by about 1970. Since then, the relative expansion of long-distance air travel has arisen largely at the expense of traffic with Ireland. This broad picture is

Table 7.16 All passenger traffic by air with the EUR12[1] countries as a percentage of all United Kingdom international air traffic

	Total	*Scheduled services*	*Charter services*
1972	64.3	58.3	73.9
1975	59.4	53.9	69.5
1980	52.7	46.1	64.2
1985	57.2	45.4	75.8
1990	56.2	49.6	69.3
1991	51.6	49.0	57.3
1992	50.9	49.0	54.9

Note: [1] Excluding Luxembourg, the Canary Islands and Madeira
Source: *UK Airports*, formerly *CAA Annual Statistics*

Table 7.17 The role of the London[1] airports in passenger traffic with the EUR12 countries[2], arrivals and departures

	Scheduled traffic			*Charter traffic*			*Total traffic*		
Year	*London (000)*	*Total (000)*	*London as %*	*London (000)*	*Total (000)*	*London as %*	*London (000)*	*Total (000)*	*London as %*
1982	10,882	12,420	87.6	4,770	12,562	38.0	15,652	24,982	62.7
1983	11,037	12,586	87.7	5,188	13,731	37.8	16,225	26,317	61.7
1984	12,097	13,927	86.9	5,810	15,708	37.0	17,907	29,635	60.4
1985	13,216	15,254	86.6	5,768	14,378	40.1	18,984	29,632	64.1
1986	13,715	15,844	86.6	6,762	18,455	36.6	20,477	34,299	59.7
1987	15,639	18,517	84.5	7,819	21,326	36.7	23,458	39,843	58.9
1988	17,244	20,975	82.2	7,688	20,624	37.3	24,932	41,599	59.9
1989	19,179	23,676	81.0	6,691	19,302	34.7	25,870	42,978	60.2
1990	21,204	26,476	80.1	5,574	16,445	33.9	26,778	42,921	62.4
1991	19,734	24,550	80.4	5,400	12,980	41.6	25,134	37,528	67.0
1992	22,432	27,705	81.0	5,704	14,530	39.3	28,136	42,235	66.6

Notes: [1] City, Gatwick, Heathrow, Stansted
[2] Excluding Luxembourg, the Canary Islands and Madeira
Source: *UK Airports*

amplified for the period since 1972 by the data in Table 7.16, showing the position of the EUR12 countries in the air traffic of the United Kingdom. The EUR12 share of traffic has fallen from 64 to 51 per cent overall. Although there has been a decline in share for both scheduled and charter services, the reduction has been least for the former. Indeed, it would appear that whereas the European countries' share has stabilized at just under 50 per cent for the scheduled services, holiday makers using charter flights are going ever-further afield.

In terms of the international origins and destinations of air travellers, there is no evidence of the post-war reorientation towards Europe that is evident in Britain's commodity trade and in the passenger traffic carried by sea. Indeed, just the opposite has been happening.

Within Britain, the airports which dominate the system are those located in and around London, and centralization arguments might lead one to expect that these airports will be gaining an increasing share of the country's traffic with EUR12. In fact, although these airports have held their share in the charter sector of the market, they have lost ground in the market for scheduled services (Table 7.17). It is these services which are of interest for business users, and quite clearly there has been a substantial relative dispersal of traffic to regional centres such as Manchester and Edinburgh, contrary to the expectations which one would derive from centralization processes.

When we consider the Channel Tunnel in Chapter 8 we will have a particular interest in traffic between London, on the one hand, and Brussels and Paris, on the other. As Table 7.18 shows, scheduled air traffic between these capital cities has indeed been expanding rapidly, but less quickly than has the total volume of scheduled traffic with EUR12 using the London airports. In other words, just as scheduled traffic with EUR12 is becoming somewhat more dispersed geographically within Britain, so also is it

Table 7.18 Passengers using scheduled services between London[1] and Brussels/Paris

	Brussels (000)	*Paris (000)*	*Brussels plus Paris as percentage of scheduled services from London to EUR12*[2]
1972	441	1,758	26.5
1975	537	1,792	25.6
1980	595	2,067	23.0
1985	717	2,421	23.7
1990	1,039	3,555	21.7
1991	969	3,286	21.5
1992	1,090	3,591	20.9

Notes: [1] City, Gatwick, Heathrow and Stansted
[2] Excluding Luxembourg, the Canary Islands and Madeira
Source: *UK Airports*, formerly *CAA Annual Statistics*

becoming less concentrated geographically in mainland Europe.

The astonishing growth in air travel that has occurred in the post-war period is apt to mask some important geographical changes which have been happening simultaneously. The main features of these changes are first the relative expansion of long-distance journeys as compared with flights to and from Europe. Furthermore, if we separate out scheduled services, for which the Channel Tunnel may realistically compete, then it is apparent that both within Britain and within mainland Europe there has been some dispersal of traffic, the net effect of which is to reduce the relative importance of traffic between London and Brussels/Paris.

CONCLUSION

The initial hypothesis was that the realignment of Britain's visible trade towards Europe, and especially towards EUR6, should imply enhanced advantages for ports (and airports) located in the South East, and that this should be reflected in the evolution of traffic patterns. Although there is clear evidence that such a shift has occurred for dry cargo, and especially for unitized traffic, exploration of the available evidence suggests very strongly that relative location may have played a rather small part in respect of freight. The implication is that in the handling of freight, distance minimization is not the determinant of trade flows; the routes selected depend at least as much on the efficiency of the ports themselves. This efficiency was differentially affected by ownership patterns and by the operation of the National Docks Labour Scheme, but recent legislative changes have for practical purposes put ports on an equal competitive footing. In the passenger sector, there is the major contrast between the great and growing concentration of seaborne traffic on Dover, now largely at the expense of neighbouring ports, and the moderate dispersal of air traffic from London, and from its services to Brussels and Paris. Altogether, the picture is quite complex and it is absolutely clear that the reorientation of Britain's economic ties towards Europe does not translate directly and simply into a concentration of traffic on the South East.

8

THE CHANNEL TUNNEL

> There is a tide in the affairs of men,
> Which, taken at the flood, leads on to fortune;
> Omitted, all the voyage of their life
> Is bound in shallows and in miseries.
> On such a full sea are we now afloat,
> And we must take the current when it serves,
> Or lose our ventures.
>
> Cassius, in *Julius Caesar*

The Channel Tunnel was officially opened on 6 May 1994 by the Queen and the President of France, thereby fulfilling a dream cherished over many generations and providing the first fixed link between Britain and France since the waning of the last ice age put them asunder. The promoters of the project have been at pains to convince us all that this major feat of engineering will have a tangible and beneficial effect on Britain as a whole; if the more enthusiastic opinions are to be believed, Britain's links with Europe will be revolutionized. If that were indeed to be the case, the impact on the whole economy would be considerable, and it is probable that the effect would be spread in a geographically uneven manner. Even if the reality is somewhat less dramatic, it is nevertheless plausible to argue that the Tunnel is significant for the relative fortunes of Britain's regions.

In this chapter, we will explore these expectations. However, because the present book was completed in 1994, any consideration of the significance of the Tunnel must be somewhat speculative. Until the Tunnel services have been operational for several years, it will not be practicable to make a full assessment. Nevertheless, it is possible to make some educated guesses which serve to identify the relevant parameters and mark out the bounds of the probable impact. In so doing, we will not dwell on the physical transformation which has been engineered just outside Folkestone and at Coquelles, near Calais in France, to create the terminal facilities. Dramatic though these changes have been, and sizeable though the employment impact is, the effects of the portal infrastructure are essentially local, albeit of great importance locally. In the present context, this local scale of impact is not of great relevance. We will, therefore, concentrate attention on the effect which the Tunnel is expected to have on the British economy as a whole and upon the relative prosperity of the regions.

If the opening of the Tunnel provides a markedly cheaper and/or more effective transport link than the ferry and air services offer for an important segment of the market, then the cost of interaction between Britain and the rest of the EU will be reduced. In this case, the effect would be analogous to a reduction in tariff and other trade barriers and should lead to economic gains similar to those which economic union is supposed to yield (Chapter 4). On the other hand, it may be the case that the improvements offered by the Tunnel – in terms of cost and quality of service – are rather limited and that the impact of the Tunnel will be quite small. Or even if the Tunnel does meet its traffic targets, it may be that capacity constraints place a clear limit to its market share. It is these and related issues on which we will focus.

To evaluate the impact of this major new link in the transport system, three sets of consideration must be taken into account.

1 The nature, quality and cost of the service offered by the Tunnel.
2 The spatial distribution of access points to the Tunnel system.
3 The competition offered by alternative means of transport, by ferry and by air.

In sum, the impact of the Tunnel cannot be assessed without a close examination of its competitive position.

In general terms, Eurotunnel claim that the Tunnel offers the traveller two major benefits: a reduction in journey times and freedom from service interruptions. The former applies primarily to traffic using the shuttle services which might otherwise use the ferries, though the Tunnel does offer through rail services between London and Brussels/Paris which are competitive with air services in terms of journey time. As for the freedom from interruption, the main benefit claimed is the absence of disruption due to bad weather, which does at times affect the ferry services. In addition, though, air services at peak holiday times are subject to delay on account of congestion in the air-space and on the runways. In respect of the cost of services, it had been expected, at the time the decision was taken to proceed with the Tunnel, that the benefits of reduced journey time and greater reliability could be offered at prices which matched the lowest available from the ferries, thereby driving average charges down.

To explore the reality of these benefits it is necessary to describe the various services which the Tunnel offers, and at what prices, as the first step to comparison with the ferry and air services. In so doing, however, it must be realized that the utility of the Tunnel is also materially affected by the effectiveness, or otherwise, of the way it is connected to the main road and rail systems on the European mainland and within Britain.

THE CHANNEL TUNNEL AS A NEW TRANSPORT LINK

The overall capacity of the tunnel is determined by the number of trains that can pass through every 24 hours, taking account of the minimum feasible spacing of trains, the mixture of train speeds and the need to maintain the track and tunnel infrastructure. At the time of opening, the planned traffic level, once all rolling stock had been delivered, was 300 trains daily in each direction with scope to increase this at least 50 per cent in due course (Eurotunnel 1993; Lord Berkeley 1994).

Three kinds of service are offered by the Tunnel:

1 The roll-on roll-off shuttle service for road vehicles, known as Le Shuttle.
2 Through rail services for passengers.
3 Through rail services for freight.

These various services became operational during the second half of 1994 and early 1995.

The shuttle services operate with three kinds of rolling stock. Double-deck enclosed wagons carry cars; single-deck enclosed wagons are used by coaches and caravans; and freight lorries travel in open-sided wagons. The configuration of shuttle trains can be varied in some degree to suit traffic conditions, by having either cars only or a mixture of cars, coaches and caravans. The transit time for Le Shuttle is 35 minutes but to this must be added the time to load and unload, and waiting time before and/or after loading. Before commercial traffic began, Eurotunnel estimated that loading and unloading would take 8 minutes, implying an *average* time per vehicle of 4 minutes at either end. On that somewhat optimistic assessment, the overall transit time would be about 45 minutes. During the summer peak season, Le Shuttle operates with departures every 15 minutes, so assuming that there is no queue, the waiting time before departure is quite short, giving an overall time from motorway to motorway of one hour. However, in the busy periods travellers must expect some queuing before departure, and if they wish to use the refreshment and off-duty sale facilities which are available at both portals, additional time must be allowed.

The driver of a car and his passengers may stay with their vehicle or walk around, but have no facilities available other than toilets in every third wagon. Coach passengers stay in their coach. In contrast, lorry drivers leave their vehicles and travel in a special compartment in which refreshments are served.

Le Shuttle operates between Folkestone and Coquelles, near Calais. In concept, it is identical to the roll-on/roll-off services offered by the ferries, except that the transit is made below the waters of the Channel instead of over its surface. In essence, therefore, the shuttle is another ferry, which

Table 8.1 Through train services from Britain to mainland Europe, other than Eurostar from London, from 1995/6

	Eurostar Beyond London daytime services		
	Travel time in hours and minutes		
	Brussels[1]	*Paris*	*Night services*
Edinburgh	8 30	8 15	London–Amsterdam
Newcastle	7 00	6 45	–Dortmund
Darlington	6 30	6 15	–Frankfurt
York	6 00	5 45	Glasgow–Brussels
Doncaster	5 30	5 00	–Paris
Newark	5 00	4 45	Swansea–Brussels
Peterborough	4 45	4 15	–Paris
Manchester Piccadilly	6 30	5 45	Plymouth–Brussels
Crewe	6 00	5 15	
Stafford	5 30	4 45	
Birmingham New Street	5 00	5 00	
Birmingham International	4 45	4 45	
Coventry	4 30	4 30	
Rugby	4 15	4 15	
Milton Keynes Central	4 00	3 45	

Note: [1] Journey times to Brussels will be reduced by 30 minutes when the Belgian high speed line opens (due in 1996)
Source: European Passenger Services 1994: 13, 19

happens to operate from slightly different points than the ferries which operate over the Dover–Calais route, which is the shortest sea crossing.

The through passenger and freight services offer something quite different. The Eurostar passenger services presently operate from Waterloo International Terminal in London to Paris and Brussels, giving travel times between city centres of 3 hours and 3 hours 15 minutes respectively. Connections at Brussels, Lille and Paris give easy access to other Continental cities. These services run at hourly intervals. From 1995 or 1996, they will be complemented by Eurostar Beyond London services, offering daytime services from fifteen points north and west of London to Brussels and Paris, and with some trains stopping at Lille. A limited number of night services, with cabins and reclining seats, will also be available (Table 8.1). At the time of writing, it was not clear what passenger services will operate from the new international station being built at Ashford, Kent.

Through freight services operate from depots in Britain (Figure 8.1) to destinations throughout Europe. Most freight trains make the Channel

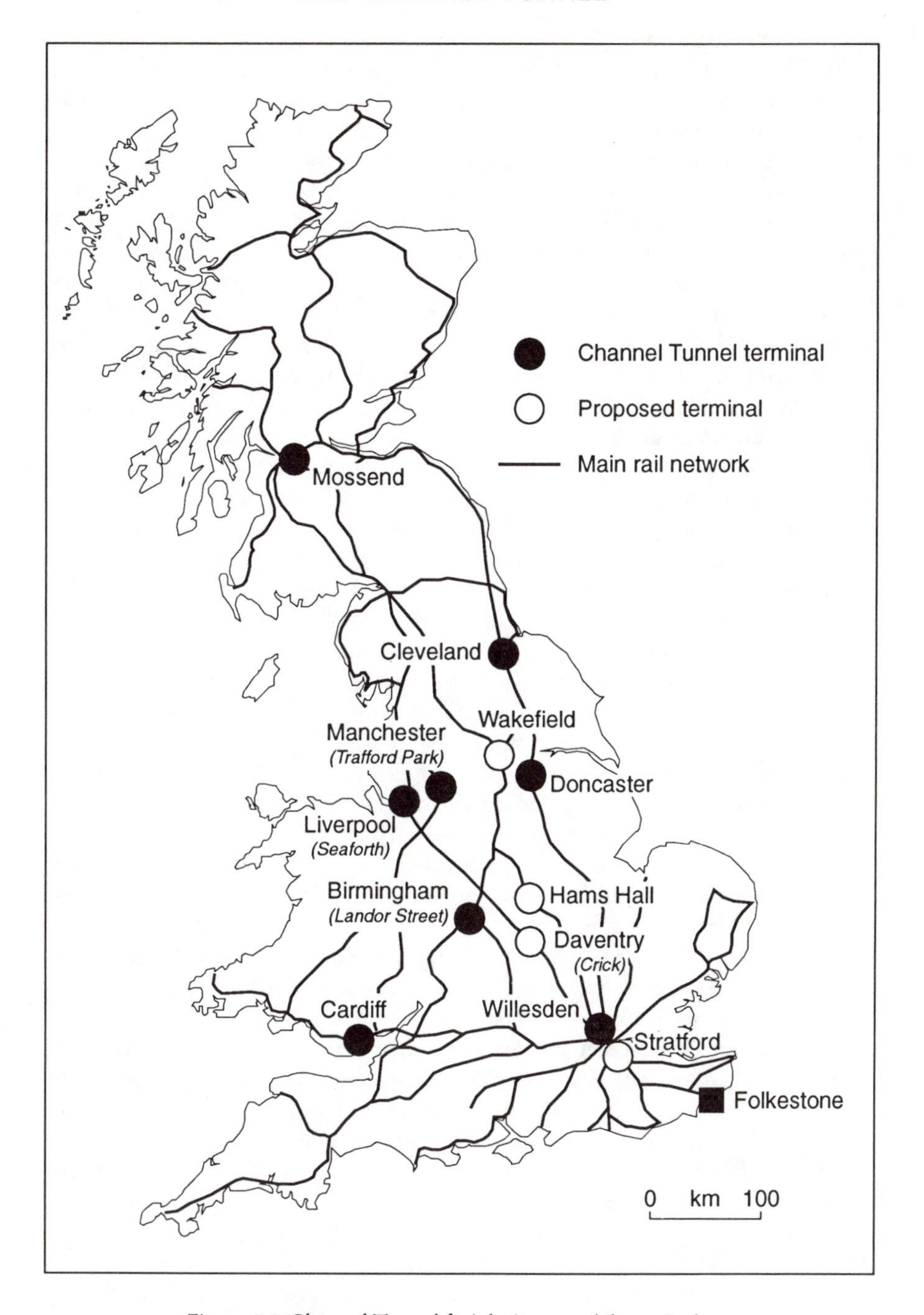

Figure 8.1 Channel Tunnel freight intermodal terminals
Source: D. Höltgen, Department of Geography, Cambridge

crossing at night, at times when passenger and shuttle demands are relatively low. As a result, freight trains leave the British depots and intermodal centres in the evening and arrive at nearby mainland destinations the following day or, for more distant places such as Barcelona, Milan and Salzburg, the day after that. To these transit times must be added the time for collection and delivery at either end of the journey.

While the number of train transits sets the upper limit for traffic through the Tunnel, the mix of this traffic is not immutably fixed and it is therefore impossible to say that the capacity is X million accompanied cars and Y million tonnes of freight, etc. In principle, and subject to the constraints imposed by the specialized nature of the rolling stock and agreements between Eurotunnel and the British and French railway companies, there can be adjustment in traffic proportions to meet the pattern of demand. That said, of the 300 trains at present passing daily in each direction, 150 are shuttle trains and 150 are through passenger and freight services. So we can make some rough calculations as to the volume of traffic that can theoretically pass through the Tunnel. If the tourist shuttle operates at 15-minute intervals and is configured to carry 120 cars and twelve coaches plus twelve caravans, then the theoretical annual capacity of traffic (summing both directions) is 8.4 million cars and 11.7 million coaches plus caravans. In addition, if all the other shuttle trains (fifty-four daily) were fully loaded with lorries, then about 1 million commercial vehicles would make the crossing, again summing for traffic in both directions (*Independent*, 6 May 1994). In practice, services in the early morning and at night will attract less custom than those operating during the period from about 09.00 to 18.00 hours, so that actual utilization will undoubtedly always be lower than the theoretical limit. Equally important, though, is the virtual certainty that at peak times the tourist shuttle will have inadequate capacity, with the result that customers will have the choice of queuing for Le Shuttle or proceeding to Dover (or Calais) for a ferry.

With thirty-five freight trains making the transit in each direction daily, each train having a capacity for 1,000 tonnes of freight, the Tunnel's theoretical annual capacity is 25.6 million tonnes in total. Each Eurostar train has seats for 794 passengers, so an hourly service from London gives a theoretical annual capacity of 13.9 million passengers travelling to and from Brussels and Paris. Initially, the Eurostar Beyond London services will provide just one train in each direction daily: along the east coast route to Edinburgh and to Manchester on the west coast. These daytime services will have a theoretical maximum capacity of 0.8 million passengers, summing traffic in both directions. The night services will offer a capacity more than double this, at about 2 million passengers.

Apart from the potential capacity constraints imposed by the Tunnel itself, an important factor which limits the scope for through rail services is the nature of the British and Continental railway systems. Two problems must be recognized. Although the gauge of the track is identical on both sides of the Channel, the Continental loading gauge permits rolling stock which is both higher and wider than can be accommodated in Britain. The effect is that although British rolling stock can operate over the European network, most European rolling stock cannot at present move over the British system – to make this possible would require the re-laying of track and the rebuilding of tunnels and bridges. As a result, dedicated rolling stock must be used. In addition, and of particular relevance to through passenger services, the electrical power systems in Britain, Belgium and France differ – indeed, there are differences within Britain as well. These differences boil down to voltage differences and electrical pick-up systems from either overhead cables or the third rail of Britain's south-eastern rail system. To avoid the delays that changing engines would cause, the units pulling passenger trains are tailor-made to cope with these system variations: expensive locomotives dedicated to a limited geographical range. Engine changes for freight trains are a less serious problem, since time is not quite so precious for freight services as it is for passengers. Nevertheless, thirty-seven freight locomotives have been purchased, designed to operate over Britain's electrified system and in France.

INTEGRATION OF THE CHANNEL TUNNEL INTO ROAD AND RAIL NETWORKS

The utility of a transport link depends rather critically on the connections it has with other parts of the transport system: the better that these links are, the more use can be made of the new link – in this case the Tunnel. To examine this question, it is manifestly necessary to consider the road and the rail connections separately.

For the Tunnel, there are excellent road connections on either side of the Channel and it can be fairly said that traffic using the shuttle services can do so as an integral part of the motorway system in both Britain and France. The main complaint in respect of road access has not been from Eurotunnel but from the competing ports and ferry services, and especially Dover. The M20 reaches the Tunnel portal just north of Folkestone, but the remaining 11 kilometres to Dover must be traversed by using the A20. Although this road is now a dual carriageway to the western edge of Dover, there is little prospect that this improvement will be carried through to the Docks. And although the A2 (which links to the M2) does give direct access to the Docks, there is again little sign that the last link of about 11 kilometres will be upgraded to dual carriageway. The main road access to Dover, now much improved, therefore passes by the competitor's front door, giving Eurotunnel a distinct advantage. In contrast, Calais is well connected to the French motorway system.

From Eurotunnel's perspective, the main problems lie with the rail links.

On the French side, there is no problem, since the TGV system reaches Coquelles and links the Tunnel into the Continental network of high-speed train services. It is on the British side that there has been a miserable saga of ideology and indecision. The Act which paved the way for the construction of the Tunnel explicitly prohibits the expenditure of public money on the Tunnel itself or indirectly through British Rail. In practice, though, it seems that governments have intervened with the banks to facilitate the funding of the construction of the Tunnel, and in underwriting the contracts entered into by the railway companies to buy slots for their trains. At the time the Bill was passing through Parliament, it was stated that no new rail link would be needed, and that the Eurostar trains would run over existing track. Hardly had the Act received the Royal assent, than pressure began to mount for the construction of a new, dedicated high-speed rail link from Folkestone to London. There are two basic reasons why such a link is needed. First, the existing rail system in Kent was already working near to capacity, so that introducing the Eurostar services would put additional pressure on a system that was under stress anyway. Second, use of the existing track implies lower speeds and reduced reliability for Eurostar than are possible on a newly engineered, dedicated, route; a fast connection to London would save between 15 and 30 minutes on the journey between London and Brussels/Paris. The public position that a new link would not be necessary probably arose from the fact that when the previous abortive attempt had been made to construct a Tunnel, it was the public outcry in Kent against the new rail link, which was integral to that scheme, which played a major part in stopping the whole project (Chisholm 1986).

During the late 1980s and early 1990s, it became abundantly clear that a new high-speed link should be built. But the government maintained that it must be constructed entirely by private finance, while simultaneously insisting on determining the route and the environmental standards to be met – mainly the extent of tunnelling. It would have been logical to propose that the cost of the cheapest option should be privately funded but that public money should be used to meet the extra costs arising from the route design standards required to protect the amenity of the area traversed by the line. In the absence of such a contribution, the private sector regarded the capital cost as too high to be warranted. So we had the farcical situation of the government willing the end but not providing the means.

This impasse was partially resolved in January 1994, when it was announced that the route would not follow the alignment preferred by British Rail, south of the River Thames until almost the last, and terminating at Kings Cross. Instead, the line will cross the Thames west of Gravesend and terminate at St Pancras. Some details of the route were still not resolved at the time of the January announcement, including the possibility of intermediate stops at Stratford, and either Rainham or Ebbfleet, and whether the international station at Ashford would be on the main line or on a spur; in

April, it was announced that the main line will pass through the new Ashford station, due to be completed in 1995. In announcing the route, one-quarter of which will be in tunnels, the government also announced that the track will be used by commuter trains. Although the speed differences between Eurostar and commuter trains poses problems for scheduling services, the decision to have mixed usage opens the way for up to £1 billion of public money to be contributed to the project. However, it is now thought that the year 2002 is the earliest that the new link will be open for traffic.

So far, the government has set its face against the idea of reconstructing a few spinal rail links north and west of London to allow Continental rolling stock to operate beyond the capital. The costs of so doing would be considerable but there is a lack of agreement on the order of magnitude. More important, there is not at present the political will to embark on such an enterprise and in any case energy is being diverted to the break-up of British Rail into a track authority and franchised rail services. The impact of these changes is not likely to be helpful to the idea of an integrated plan for rail services in Britain and will almost certainly delay the full integration of the Tunnel and the British rail system into the Continental network. This in turn means that the full benefits of the Tunnel for through passenger services will not be realized for a long time.

The position is somewhat more complex in respect of freight services. The basic problem relates not to the width of rail wagons but to their height. The basic reason for this lies in the fact that conventional rail wagons now carry very little freight – mainly coal, cement and other bulk commodities. There is, therefore, very little need for conventional rolling stock to enter Britain. To an increasing extent, rail freight is carried in units which at some point travel over the road system – complete lorries, swap bodies and containers. These are subject to regulations throughout the EU which require vehicles and their loads to be narrower than the loading gauge for British railways requires. Therefore, there is no problem on account of the width of loads with respect to the railway freight traffic which passes through the Tunnel.

Containers are robust boxes which can be stacked. In contrast, a swap body is a container which is built to lighter (road) specifications and which, as a consequence, cannot be stacked. In both cases, containers or swap bodies are carried on a flat vehicle, whether a road vehicle or a rail wagon, without being stacked. When carried by rail, the overall height of bridges and tunnels gives adequate clearance, but their arched construction means that, in some cases, the outside edge of the container or swap body would foul the structure. The mismatch is a matter of only a few inches vertically and the solution is relatively straightforward; either the track can be lowered by the small amount that is necessary, or low-platform EuroFret wagons can be used. British Rail is in fact modifying key routes of the network to accommodate containers and swap bodies (Dick *et al.* 1993).

Unfortunately, containers account for only 16 per cent of cross-Channel unitized freight, and rather few British hauliers have as yet invested in swap bodies. As a consequence, the improvements which British Rail is making tap into a small segment of the market for freight traffic with the Continent, which is dominated by trucks, more than four-fifths by volume being handled in this manner. The majority of these trucks are articulated – that is, the trailer with the load can be detached from the tractor unit. The trailer itself is an integral structure of chassis, wheels and cargo container. To carry either the trailer, or the complete assemblage of tractor and trailer, implies an overall height in excess of that required for containers and swap bodies. If it is just the trailer which is carried (a system known as 'Piggyback'), then a partial solution is to use a wagon with a low-slung pocket for the wheels and chassis of the trailer, as happens elsewhere in Europe. Even then, the height exceeds the height of a container or swap body. If the whole vehicle is to be transported, it must be carried on a flat wagon and an even greater height is required.

If British Rail could carry trailers piggyback, Dick *et al.* (1993) believe that the railways could capture approaching one-third of the unitized traffic which crosses the Channel. In other words, instead of the 9 per cent market share which Dick *et al.* estimate British Rail would win under present proposals, the share could be about 32 per cent. To achieve this share would require investment to modify bridges and tunnels, at a cost which Dick *et al.* estimate to be about £300 million, but which British Rail consider would be substantially greater. To address the issue of loading gauge and Piggyback, a grouping of rail freight operators, track authorities, and strategic local authorities, working as The Piggyback Consortium, undertook in 1993 a national study with the support of the European Commission. The study concluded that Piggyback services could be commercially viable, especially given the indifference of UK industry to the use of swap bodies and its preference for the Piggyback concept. The Consortium is currently working on the development of Piggyback equipment and loading gauge improvements to rail routes in Britain (Mowatt 1994).

THE TUNNEL AND PASSENGER FERRY SERVICES

The ferry services in the most direct competition with the shuttle facilities offered by the Tunnel operate between Dover and Calais. Two companies – P. & O. and Stena-Sealink – operate conventional ferries, and Hoverspeed offer hovercraft services. Prior to the opening of the Tunnel, the Hoverspeed services accounted for about 10 per cent of the traffic, notwithstanding that the crossing times, at 35 minutes, are much shorter than the ferries can offer. The pre-Tunnel market was dominated by the two ferry companies, which offer crossing times of 90 minutes in the case of Stena-Sealink and 75 minutes on the P. & O. services. For both companies, the check-in time is 20 minutes before departure.

Both ferry companies have invested in refurbishing vessels, P. & O. have also invested in new vessels, and the dock facilities at Dover and Calais have been improved to facilitate speedy handling of traffic and foot passengers. So how does the ferry option compare with the Tunnel? When Stena-Sealink announced their schedule and tariffs for 1994, they increased the number of sailings from the 1993 level of twenty-two to twenty-five each day, matching the number provided by P. & O.; the ferries therefore offer fifty crossings each way every day. If these crossings were spread uniformly through the 24 hours, the interval between sailings would be 28.8 minutes. However, the companies are prevented from setting their sailing schedules in consultation with each other, and in any case offer less frequent services during the night than the day. Some of their sailings depart at exactly the same time. Averaged over the 24 hours, the interval between sailings is in fact 32 minutes, being a little more than this until 08.00 hours and slightly less throughout the remainder of the day. Overall, we can say that the frequency of the ferries is half that of Le Shuttle, with 30 minutes separating sailings compared with 15 minutes for the shuttle. However, in almost one-third of the cases, the interval between ferries is only 15 minutes. This overall position is little affected by the inclusion of the Hoverspeed services, since most of their twelve daily crossings coincide with ferry departure times. The minimum road to road time for a ferry crossing is about 100 minutes, some 40 minutes longer than the minimum by Le Shuttle. However, P. & O. reckon that the average actual time differential is less than this: taking account of waiting times experienced for Le Shuttle, and the fact that very large numbers of vehicles travel on an earlier ferry than the one for which they are booked, P. & O. estimate that the real differential is only 20 minutes.

While the structure of services actually offered by the ferry companies and Eurotunnel is reasonably stable in the short run of a single year, the prices charged by the operators are apt to vary at short notice. Furthermore, the standard published fares, on which comparisons can most readily be made, account for only a proportion of passenger ticket revenue; in the case of P. & O., half the ticket revenue arises from traffic benefiting from discounts of one kind or another. There are, therefore, considerable difficulties in making valid comparisons. In January 1994, Stena-Sealink announced fares which, the company estimated, represented an overall increase over the previous year of 6 per cent (*Independent*, 6 January 1994). Shortly thereafter, Eurotunnel, expecting to offer services in May, announced its own fares. By April, both Stena-Sealink and P. & O. had published the third edition of their brochures, and in fact Eurotunnel's services were not started until considerably later than had been expected. It is quite clear that, at the time of writing, the price competition between Eurotunnel and the ferries was in its early phase. Indeed, to everyone's surprise, in May Stena-Sealink announced a 20 per cent reduction on selected sailings (*Independent*, 25 May 1994), before Eurotunnel had opened Le Shuttle for normal services. Nevertheless,

Table 8.2 Return fares for accompanied cars, in £ sterling, 1994

		Summer		
	Winter	*Off-peak*	*Peak*	*Peak weekend*
Eurotunnel (unlimited passengers)				
Standard return	220	260	280	310
5-day return	130	160	–	–
P. & O. (up to eight passengers)				
Standard return	139	190	220	320
5-day return	76	124	130	175
Stena-Sealink (up to four passengers)				
Standard return	126	188	220	320
5-day return	60	124	126	174
Hoverspeed (up to four passengers)				
Standard return	138	270	309	324
5-day return	75	160	174	189

Sources: Published travel brochures: Eurotunnel and Hoverspeed, 1st edition; Stena-Sealink and P. & O., 3rd edition. All as of 19 April 1994

some important characteristics of the competition can be discerned.

Table 8.2 shows that the two ferry companies have standard fares that are very similar to each other, and in general rather cheaper than Hoverspeed and Eurotunnel. However, this table explicitly reveals three further facts but hides a fourth. During the period of peak travel demand, Eurotunnel's standard rates closely match those of the three surface companies, being just marginally lower. For practical purposes, there is little to choose between operators regarding the price of a standard return. However, it will be noted, second, that Eurotunnel do not offer a cheap five-day return in the peak periods. In the third place, though, Eurotunnel's charges in the off-peak summer period and in winter are substantially higher than those charged by their competitors, and more especially in winter, when they are between 60 and 70 per cent higher than by ferry and about 100 per cent higher in the case of five-day return fares. The reasoning behind this differential is that, especially in the winter season, bad weather may disrupt the services offered by the ferries and Hoverspeed; Eurotunnel calculate that they can charge a premium for reliability during these months, a claim hotly disputed by the ferry companies. The fourth feature of Eurotunnel's pricing is not revealed by Table 8.2. Although Eurotunnel vary their charges by time of year, they make no such variation by time of day, unlike P. & O. and Hoverspeed. Thus, even on the peak weekends it is possible to travel by ferry for £190 return on some of the very early morning and late evening services. Other than at the peak weekends, if one is willing to travel in the early morning or late evening, it is possible to do so at the winter rate of £139. As a consequence, for those who are willing to cross at the unpopular times during the summer it is possible to do so at between about 50 per cent and 70 per cent of the standard cost by Eurotunnel, a differential similar to the differential already noted for the winter services.

Traditionally, all the ferry operators have used a tariff which differentiates charges by time of day. Although Stena-Sealink initiated their 1994 prices on that basis, they subsequently came in line with Eurotunnel. However, the probability is that these two companies will revert to the well-tested principle of differentiating by time of day, as is the continuing practice of P. & O. and Hoverspeed, to maximize use of capacity.

At any rate, in the opening rounds of competition, it is quite clear that far from matching the lowest discounted fares by ferry, Eurotunnel is in fact charging a premium price throughout the year at certain times of the day, or throughout the 24 hours in winter. This fact, which reflects the cost over-run in constructing the Tunnel and the delay in starting services, means that competition is much less severe for the ferries than they had feared in the mid-1980s.

Although the ferries take longer than Le Shuttle, they exploit this fact by offering a variety of options for eating and drinking, amusements, shops and duty-free facilities, although the latter will in due course cease to exist as EU regulations are harmonized. To combat the advantage of speed offered by Eurotunnel, the ferries provide a break from driving, which for most people can be an enjoyable experience, and the possibility of making substantial cost savings. On the face of it, the competitive situation with Eurotunnel looks more favourable to the ferries than many had expected about a decade ago (but see Chisholm 1986), and the probability has to be that Eurotunnel will be forced to reduce at least some of its fares.

In Chapter 7 we examined the growth of traffic between Britain and Europe, noting the importance of the ports along the Thames, Kent, Sussex and Hampshire coast for accompanied cars. If we exclude services to Ireland and Northern Ireland, then in 1993 a total 5.5 million cars were carried on international and coastal services through the ports of Great Britain. Of that total, 3.1 million used Dover, and an additional 1.6 million the other ports from the Thames to Hampshire (Dunlop 1994). These figures may be compared with the theoretical capacity of the Tunnel (if Le Shuttle operates throughout the year with a service interval of 15 minutes) of about 8 million cars. Theoretically, therefore, Le Shuttle could handle Great Britain's entire existing traffic in accompanied cars. In practice, Eurotunnel operates at less than full capacity for most of the time but clearly has the ability to make serious inroads into Dover's traffic in particular. Eurotunnel itself expects to carry about 8 million car passengers (not cars) by 1996 or about one-half of the projected cross-Channel traffic (*Financial Times*, 12 January 1994). If the traffic does divide in this manner, there will be considerable financial pressure on all the operators, but a substantial continuing role for Dover and other ports.

THE TUNNEL AND AIR SERVICES

A three-hour Eurostar rail journey from Waterloo to Paris, and just 15 minutes longer to Brussels, is an attractive alternative to flying in terms of the time from one city centre to another. The lowest initial inter-capital fare was slightly higher than the cheapest air fare (£95 return compared with £83), but standard and first-class fares were substantially lower than air travel. The Eurostar services are competing most strongly for the business market (*Independent*, 18 October 1994).

Each Eurostar train is designed to seat 794 passengers. An hourly service gives a theoretical capacity of just over 38,000 passengers daily travelling between London and the two European capitals, or 13.9 million each year. To this capacity must be added the services which will operate from points north and west of London, but these will not have the same competitive advantage in time compared with air services to Brussels and Paris that is available from Waterloo. If we reckon that the system capacity with hourly services from London to the two capitals is about 14 million passengers, there clearly is the potential for serious competition with the air services.

Airbus calculate that, on short-haul European routes, trains will handle about half the traffic if the doorstep-to-doorstep time by rail and air is the same (*Economist*, 12 September 1992). Table 7.18 shows that the inter-capital market for scheduled air services has been growing steadily but only totalled 4.7 million in 1992. If the Airbus estimate is near the mark, about 2.35 million passengers would transfer from air to rail, plus half the incremental traffic. Beyond Brussels and Paris daytime rail services steadily lose their advantage in time, and transfers to the rail system for these further destinations will be much less than in the case of the inter-capital services.

These considerations suggest that Eurotunnel's 1991 expectation that in the year of opening it would handle 15 million passengers, excluding shuttle passengers, was wildly optimistic (Eurotunnel 1991: 4). It would appear that the Civil Aviation Authority, in preparing advice to the Secretary of State for Transport on the need for additional airport capacity in the London area, took a more realistic view of passenger traffic through the Tunnel (Civil Aviation Authority 1990: 21). They estimated that, in 1995, 3.3 million passengers would be 'lost' to rail services through the Tunnel, rising to 7.0 million in 2005, equivalent to 4 and 6 per cent respectively of all passenger movements through the London airports. Even at this relatively modest level of traffic diversion, the impact on air services to Brussels and Paris would be substantial.

Two future developments may have a significant impact on the relative attractiveness of rail and air services between London and Brussels/Paris. In 1997, a new rail link is due to be opened, providing access from Paddington to Heathrow. Trains will run every 15 minutes, offering a journey time of only 16 minutes. Furthermore, if in due course the London crossrail link is constructed, passengers from east of London and Liverpool Street will have direct access to Paddington and thence to Heathrow. As a result, Heathrow will become significantly more accessible and journey times will be reduced by as much as 30 minutes or even more. Reflecting this substantial time saving, it is expected that the fare from Paddington will be about four times the fare on the Piccadilly Line – at 1994 prices a single fare of £9 compared with £2 (*Guardian*, 26 January 1994). The effect should be to enhance the attractions of air services from Heathrow relative to the Tunnel.

The second issue concerns the profitability of the European airlines, the scale of public subsidy and the prices which customers are charged. In general terms, the airline capacity of Europe is reckoned to be excessive, largely because national pride puts a high premium on maintaining the national flag carriers. For all the talk about deregulation, relatively little has in fact been done, and it remains the case that air travel in Europe is expensive compared with elsewhere in the world. The scale of the problem is indicated by Table 8.3. Of course, these figures refer to the worldwide operations of the companies, not just to services between Britain and Europe. However, the potential impact of changes in subsidy regime and the pattern of service regulation is clearly considerable. Most commentators expect that the pressure will mount for the rationalization of services and the elimination of

Table 8.3 European airline profits, losses and subsidies, £ million

Airline	*Pre-tax profits and losses 1992*	*Subsidies approved by the EU since 1991*	*New subsidies under consideration, February 1994*
British Airways	199		
Sabena	8	667	
Alitalia	–8		
Aer Lingus	–130	167	
TAP	–133	133	667
Olympic	–150		
Lufthansa	–166		
KLM	–212		
Iberia	–226	667	
Air France	–411	800	167

Source: *Daily Telegraph*, 2 February 1994

surplus capacity, and that moves to reduce regulation will serve to drive fares downwards. The availability of rail services through the Tunnel will add to the competitive pressures, but it is unlikely of itself to be the major factor in any changes that occur. Clearly, though, in considering the competition between the Tunnel and the airlines, attention must be paid to developments in the airline industry.

The most immediate impact of the Tunnel on air services will be on the London–Brussels/Paris routes, where it is clear that significant inroads will be made. The airlines are not likely to be unduly bothered, since if they reduce the number of flights offered they will be able to redeploy their landing slots at the London airports for other routes. The overall impact on their operations will be small. On the other hand, it does seem a little difficult to visualize the Eurostar services achieving their capacity passenger loads for quite some years. Much may therefore depend on the build-up of daytime traffic from points north and west of London; and of the night services, for which journey time is less important than the fact that it is possible to travel at the same time as sleeping.

THE TUNNEL AND FREIGHT SERVICES

Information concerning the charges for freight services, especially through rail, is much less readily available than is information on passenger services. If we take Eurotunnel's own forecasts at face value, then in 1991 they were expecting to handle 8.5 million tonnes of shuttle freight and 6.9 million tonnes of through rail freight on a full-year basis in 1993, rising to 15.3 and 12.6 million tonnes respectively by 2003 (Eurotunnel 1991: 4). A year later, no estimates were given for the opening year (1994), but the 2003 forecasts had been scaled back to 14.7 and 11.7 million tonnes for shuttle and rail freight, implying some reduction in the volume expected in the early years of operation.

We can make an initial assessment of these expectations in the following manner. All of the freight handled by Eurotunnel – shuttle freight and through rail freight – can be assumed to be unitized, and therefore competing in a market which in 1992 amounted to 77 million tonnes (Table 7.10). If Eurotunnel handle about 15 million tonnes per annum in the early years, that will be a substantial but not a dominant share of the market. At that traffic level though, the Tunnel would have considerable spare capacity for both shuttle and through freight services. However, to achieve even 15 million tonnes implies substantial inroads into the traffic of south-eastern ports, and in particular Dover, which handles about 13 million tonnes annually. Since it is unlikely that Dover will lose all its roll-on roll-off traffic, it seems probable that Eurotunnel and the rail operators will have to work hard to achieve their market aims. One reason why the time saving offered by the Tunnel may not be particularly attractive to road hauliers is the fact that drivers must, by law, have rest periods, and the ferries offer a much more useful rest period than does Le Shuttle. On the other hand, unitized freight is the most rapidly growing sector of the freight business, so that much of the short-term excess capacity is likely to be used in the relatively near future.

CONCLUSION

The ferry companies operating the short sea routes (not just Dover–Calais) have a collective investment of about £1 billion. The Tunnel represents a further £10 billion. Since there is no possibility of a tenfold increase in traffic in the foreseeable future, it follows that there will be intense pressure on all the companies in the fight to maintain profit levels, and the virtual certainty that they will not all be able to do so. The pre-Tunnel return on capital earned by the industry is bound to fall, as also is the real cost of travel for passenger and freight customers. But paradoxically, it is not in the interests of the ferry operators for Eurotunnel to become bankrupt, since it is certain that the assets would be operated but, following bankruptcy, with a much-reduced capital liability, and hence the ability to cut prices. From the point of view of the ferries, a major key to the future will be whether, or when, they are permitted to co-operate in setting their schedule of sailings, or whether Stena-Sealink and P. & O. join forces as one company.

At the national level, the main impact of the Tunnel is to transfer traffic from one mode to another, with only a modest effect in generating new traffic. The existence of the Tunnel puts downward pressures on prices, and in this sense brings Britain a little nearer to Europe. This is equivalent to a small reduction in tariff or other trade barriers. Given the evidence reviewed in Chapters 4 and 6, it seems unlikely that this will have a discernible impact on the aggregate performance of the British economy.

The Tunnel may, however, have an important impact on the relative fortunes of Britain's regions (Button 1990, 1994; Chisholm 1986; Gibb 1994; Gibb *et al.* 1992; Holliday *et al.* 1991, Tolley and Turton 1987; Vickerman 1989; Vickerman and Flowerdew 1990). The impact at the Tunnel portal has been huge, and the operation of the Tunnel is likely to have a significant impact on Dover. However, these impacts relate primarily to the transfer, real and potential, between alternative forms of transport which are using Kent as a routeway. To the extent that the Tunnel generates more traffic, the pressure on Kent's transport systems will be increased. As partial compensation, Ashford will be available as a point at which Eurostar services can be accessed, and, when the fast rail link is completed, there will be some improvement in rail commuter services to London for a number of stations. Overall, Kent gains little from increased accessibility, and loses considerably from the pressure of transit traffic.

London clearly gains form the existence of an additional means of travel to mainland Europe, and this will serve to enhance its attractions for business and for workers relative to the rest of the country. But the magnitude of that differential gain depends on the quality of services offered to destinations beyond London, which are set to improve from the mid-1990s onwards. So far as freight is concerned, rail services offer distinct advantages over longer distances compared with road services, and this should therefore mean some relative gain for the north and west of the country.

There is little doubt that come the 1980s the time was ripe to build the Tunnel. However, whether its construction will lead on to fortune remains a moot point. The Tunnel provides a major new element in Britain's links with Europe, which brings her somewhat closer to the mainland than previously was the case, but which does not have the capability to effect a revolutionary change. Since being on the edge of Europe does not matter that much anyway, being a little closer is not going to make a huge difference to the national economy or even, probably, to differential regional fortunes.

9

CONCLUSION

No man is an *Island*, entire of it self
John Donne, *Meditation XVII*

In Chapter 1, we reviewed some of the ideas and literature which suggest that Britain's location on the edge of mainland Europe puts her at a permanent disadvantage in economic terms, and that further integration within the EU should be resisted. Our exploration of these ideas has taken us into some quite disparate literature and data that reveal several conclusions that really can be accepted without much reasonable doubt. The dramatic reorientation of Britain's trade towards Europe in the post-war period is quite clearly but part of a general trend for the industrialized nations to trade with each other rather than with the suppliers of primary produce. In Britain's case, this reorientation has been substantially more dramatic than has been the experience of the other European countries, reflecting, at least in part, the rapid decline of Commonwealth links. Equally important, though, Britain's trade realignment had begun before the EU was formed in 1958 and has continued steadily irrespective of whether she was or was not a member. And the fact that there has been a deterioration in the balance of visible trade with the EU matches the fact that the same adverse trend has been evident in trade with the rest of the world.

Examination of the role of distance in international trade, of transport costs in Britain, of the impact of economic integration and of the supposed consequences of cumulative growth provides very little evidence to support the contention that Britain's location peripheral to Europe results in any measurable effect that its adverse. That conclusion is based on commodity production and trade. The international service sector, it is argued, is less sensitive to distance costs than is the commodity sector, so it follows that if a peripheral location is of little moment for Britain in commodity production and trade, it will be even less significant in the commerce associated with the service industries.

It is quite clear that the conventional wisdom which holds that Britain's location on the edge of Europe is a disadvantage is in fact mistaken. In other words, if the British economy has problems, they cannot in modern times be ascribed to the accident of location on the face of the Earth. If we perceive that there are problems – and most observators would agree that there are – then we do not have the scapegoat of location to blame. We must look elsewhere, both for the nature of the problems and for the solutions. By the same token, of course, to propose the maintenance or even accentuation of barriers to trade and commerce with Europe would profit us nothing.

The future of Britain is inextricably bound up with Europe and will so remain for the foreseeable future, and indeed the links will become stronger rather than weaker. If that means that some of our institutions must change, that Parliament will become less 'sovereign', that is an inescapable consequence of changes which go far beyond the formation and consolidation of the EU. It really is the case that we are no longer able to act wholly independently, sufficient unto ourselves. In the domain which we have reviewed in this book, which is of course only part of the wider economic and political scene, it seems entirely reasonable to propose with the *Independent* (29 November 1991) that we have nothing to fear but fear itself.

BIBLIOGRAPHY

Altham, P., M. Chisholm and A. Cliff (forthcoming) 'Distance and the pattern of international trade', offered to the *Journal of the Royal Statistical Society*, Series A.

Anderson, K. and R. Blackhurst, (eds) (1993) *Regional Integration and the Global Trading System*, Hemel Hempstead: Harvester Wheatsheaf.

Artis, M.J. (ed.) (1986) *Prest and Coppock's The UK Economy. A manual of applied economics* (11th edn), London: Weidenfeld and Nicolson.

Balassa, B. (1962) *The Theory of Economic Integration*, London: Allen & Unwin.

——— (1974) 'Trade creation and trade diversion in the Common Market: an appraisal of the evidence', *Manchester School of Economic and Social Studies* 42: 93–135.

——— (1986) 'The determinants of intra-industry specialization in United States trade', *Oxford Economic Papers* 38: 220–33.

——— (1989) *Comparative Advantage, Trade Policy and Economic Development*, Hemel Hempstead: Harvester Wheatsheaf.

Bassett, K. (1993) 'British port privatization and its impact on the port of Bristol', *Journal of Transport Geography* 1: 255–67.

——— and A. Hoare (1984) 'Bristol and the saga of Royal Portbury: a case study in local politics and municipal enterprise', *Political Geography Quarterly* 3: 223–50.

Batchelor, R.A., R.L. Major and A.D. Morgan (1980) *Industrialisation and the Basis for Trade*, Cambridge: Cambridge University Press.

Batten, D.F. (1983) *Spatial Analysis of Interacting Economies. The role of entropy and information theory in spatial input–output modeling*, Boston: Kluwer-Nijhoff.

Batten, D.F. and D.E. Boyce (1986) 'Spatial interaction, transportation, and interregional commodity flow models', in P. Nijkamp (ed.) *Handbook of Regional and Urban Economics*, Vol. 1, *Regional Economics*, Amsterdam: North Holland, 357–406.

Bayliss, B.T. and S.L. Edwards (1970) *Industrial Demand for Transport*, London: HMSO.

Beckerman, W. (1956) 'Distance and the pattern of intra-European trade', *Review of Economics and Statistics* 38: 31–40.

Beenstock, M. and P. Warburton (1983) 'Long-term trends in economic openness in the United Kingdom and the United States', *Oxford Economic Papers* 35: 130–5.

Begg, I. (1990) 'The Single European Market and the UK regions', in G. Cameron, B. Moore, D. Nicholls, J. Rhodes and P. Tyler (eds) *Cambridge Regional Economic Review. The economic outlook for the regions and counties of the United Kingdom in the 1990s*, Cambridge: Cambridge Economic Consultants and Department of Land Economy (University of Cambridge), 89–104.

——— and D. Mayes (1993) 'Cohesion, convergence and economic and monetary union in Europe', *Regional Studies* 27: 149–55.

Bhagwati, J. (1991) *The World Trading System at Risk*, Hemel Hempstead: Harvester Wheatsheaf.

Black, W. (1972) 'Inter-regional commodity flows: some experiments with the gravity model', *Journal of Regional Science* 12: 107–18.

Blackaby, F. (ed.) (1978) *De-industrialisation*, London: Heinemann.

Borts, G.H. and J.L. Stein (1964) *Economic Growth in a Free Market*, New York: Columbia University Press.

Brams, S.J. (1966) 'Transaction flows in the international system', *American Political Science Review* 60: 880–98.

Brenton, P.A. and L.A. Winters (1992) 'Bilateral trade elasticities for exploring the effects of "1992"' in L.A. Winters (ed.) *Trade Flows and Trade Policy after '1992'*, Cambridge: Cambridge University Press, 226–85.

Bröcker, J. (1988) 'Interregional trade and economic integration. A partial equilibrium approach', *Regional Science and Urban Economics* 18: 261–81.

——— and K. Peschel (1988) 'Trade', in W. Molle and R. Cappellin (eds) *Regional Impact of Community Policies in Europe*, Aldershot: Avebury, 127–51.

Brown, A.J. (1969) 'Surveys of applied economics: regional economics, with special reference to the United Kingdom', *Economic Journal* LXXIX: 759–96.

——— (1972) *The Framework of Regional Economics in the United Kingdom*, Cambridge: Cambridge University Press.

Brown, C.S.F. and T.D. Sheriff (1978) 'De-industrialisation: a background paper', in F. Blackaby (ed.) *De-industrialisation*, London: Heinemann, 233–62.

Bryan, I.A. (1974) 'The effect of ocean transport costs on the demand for some Canadian exports', *Weltwirtschaftliches Archiv* 110: 642–62.

Button, K. (1990) 'The Channel Tunnel – the economic implications for the South East of England', *Geographical Journal* 156: 187–99.

——— (1994) 'The Channel Tunnel and the economy of southeast England', *Applied Geography* 14: 107–21.

Caballero, R.J. and R.K. Lyons (1991) 'External effects of Europe's integration', in L.A. Winters and A. Venables (eds) *European Integration: trade and industry*, Cambridge: Cambridge University Press, 34–51.

Cameron, G.C. and B.D. Clark (1966) 'Industrial movement and the regional problem', *Occasional Papers* No. 5, University of Glasgow Social and Economic Studies, Edinburgh: Oliver & Boyd.

——— and G.L. Reid (1966) 'Scottish economic planning and the attraction of industry', *Occasional Papers* No. 6, University of Glasgow Social and Economic Studies, Edinburgh: Oliver & Boyd.

Cecchini, P. (1988) *The European Challenge 1992. The benefits of a single market*, Aldershot: Wildwood.

Chisholm, M. (1959) 'Shipping costs and the terms of trade: Australia and New Zealand', *Applied Statistics* VIII: 196–201.

——— (1962) *Rural Settlement and Land Use: An essay in location*, London: Hutchinson.

——— (1963) 'Tendencies in agricultural specialization and regional concentration of industry', *Papers*, Regional Science Association, X: 157–62.

——— (1964) 'Must we all live in southeast England? The location of new employment', *Geography* XLIX: 1–14.

——— (1966) *Geography and Economics*, London: Bell.

——— (1971) 'Freight transport costs, industrial location and regional development', in M. Chisholm and G. Manners (eds) *Spatial Policy Problems of the British Economy*, Cambridge: Cambridge University Press, 213–44.

——— (1979) *Rural Settlement and Land Use. An essay in location* (3rd edn) London: Hutchinson.

——— (1985a) 'De-industrialisation and British regional policy', *Regional Studies* 19: 301–13.

——— (1985b) 'Accessibility and regional development in Britain: some questions arising from data on freight flows', *Environment and Planning* A 17: 963–80.

——— (1986) 'The impact of the Channel Tunnel on the regions of Britain and Europe', *Geographical Journal* 152: 314–53.

——— (1987) 'Regional variations in transport costs in Britain, with special reference to Scotland', Institute of British Geographers, *Transactions* 12: 303–14.

——— (1990a) 'The increasing separation of production and consumption', in B.L. Turner, W.C. Clark, R.W. Kates, J.F. Richards, J.T, Mathews and W.B. Meyer (eds) *The Earth as Transformed by Human Action. Global and regional changes in the biosphere over the past 300 years*, Cambridge: Cambridge University Press, 87–101.

——— (1990b) *Regions in Recession and Resurgence*, London: Unwin Hyman.

——— (1992) 'Britain, the European Community, and the centralisation of production: theory and evidence, freight movements', *Environment and Planning* A, 24: 551–70.

——— (1995) 'Britain at the heart of Europe?', in A.D. Cliff, A.G. Hoare, N. Thrift and P. Gould (eds) *The Unity of Geography*, Institute of British Geographers Special Publication Series, Oxford: Basil Blackwell.

Chisholm, M. and P. O'Sullivan (1973) *Freight Flows and Spatial Aspects of the British Economy*, Cambridge: Cambridge University Press.

Civil Aviation Authority (1990) *Traffic Distribution Policy and Airport and Airspace Capacity: the next 15 years. Advice to the Secretary of State for Transport*, CAP 570, London: CAA.

Clark, C. (1966) 'Industrial location and economic potential', *Lloyds Bank Review*, October: 1–17.

——— (1967) *Population Growth and Land Use*, London: Macmillan.

———, R. Wilson and J. Bradley (1969) 'Industrial location and economic potential in Western Europe', *Regional Studies* 3: 197–212.

Cliff, A.D., R.L. Martin and S.K. Ord (1976) 'A reply to the final comment', *Regional Studies* 10: 341–2.

Commerzbank (1992) 'Intra-industry trade drives growth in Europe', *Economist*, 18 July, p. 70.

Commission of the European Communities (1989) 'International trade of the European Community', *European Economy* 39.

——— (1990) 'One market, one money. An evaluation of the potential benefits and costs of forming an economic and monetary union', *European Economy* 44.

——— (Directorate-General for Regional Policy) (1991) *The Regions in the 1990s. Fourth periodic report on the social and economic situation and development of the regions of the Community*, Luxembourg: Office for Official Publications of the European Communities.

——— (1993) 'Report on progress with regard to economic and monetary convergence and with the implementation of community law concerning the internal market', *European Economy* 55: 43–104.

Committee for the Study of Economic and Monetary Union (1989) *Report on Economic and Monetary Union in the European Community* (with papers submitted to the Committee), Luxembourg: Office for Official Publications of the European Communities.

Corbridge, S. and J. Agnew (1991) 'The US trade and budget deficits in global perspective: an essay in geopolitical-economy', *Environment and Planning* D, 9: 71–90.

Crouch, C. and D. Marquand (eds) (1992) *Towards Greater Europe? A continent without an iron curtain*, Oxford: Blackwell.

Curry, L. (1972) 'A spatial analysis of gravity flows', *Regional Studies* 6: 131–47.

Curwen, P. (ed.) (1990) *Understanding the UK Economy*, Basingstoke: Macmillan.

Delors, J. (1989) 'Regional implications of economic and monetary integration', in Committee for the Study of Economic and Monetary Union, *Report on Economic and Monetary Union in the European Community*, Luxembourg: Office for Official Publications of the European Communities, 81–9.

Department of Transport (1993) 'Origins, destinations and transport of UK international trade 1991', *Statistics Bulletin* 93: 32, London: Department of Transport.

Dick, A., D. Baker and M. Garrett (1993) *Moving International Freight from Road to Rail. The loading gauge issue*, London: The Channel Tunnel Group.

Dicken, P. (1986) *Global Shift. Industrial change in a turbulent world*, London: Paul Chapman.

Dunford, M. (1993) 'Regional disparities in the European Community: evidence from the REG10 databank', *Regional Studies* 27: 727–43.

——— (1994) 'Winners and losers: the new map of economic inequality in the European Union', *European Urban and Regional Studies* 1, 95–114.

Dunlop, G. (1994) Personal communication.

Edwards, S.L. (1975) Regional variations in freight cost', *Journal of Transport Economics and Policy* IX: 1–12.

El-Agraa, A.M. (1984) 'Has membership of the European Communities been a disaster for Britain?', *Applied Economics* 16: 299–315.

——— (1989) *The Theory and Measurement of International Economic Integration*, Basingstoke: Macmillan.

——— (1990a) 'The theory of economic integration', in A.M. El-Agraa (ed.) *Economics of the European Community* (3rd edn), Hemel Hempstead: Philip Allan, 79–96.

——— (1990b) 'General introduction', in A.M. El-Agraa (ed.) *Economics of the European Community* (3rd edn), Hemel Hempstead: Philip Allan, 1–17.

Emerson, M., M. Aujean, M. Catinat, P. Goybet and A. Jacquemin (1988) *The Economics of 1992. The E.C. Commission's assessment of the economic effects of completing the internal market*, Oxford: Oxford University Press.

Erickson, R.A. and D.J. Hayward (1991) 'The international flows of industrial exports from U.S. regions', Association of American Geographers, *Annals* 81: 371–90.

European Passenger Services (1994) *International Train Services Fact Sheets*, London: EPS.

Eurotunnel (1991) *Interim Report 1991*, Folkestone: Eurotunnel.

——— (1992) *Interim Report 1992*, Folkestone: Eurotunnel.

——— (1993) *Eurotunnel Information Paper: transport infrastructure*, Folkestone: Eurotunnel.

Fetherston, M., B. Moore and J. Rhodes (1979) 'EEC membership and UK trade in manufactures', *Cambridge Journal of Economics* 3: 399–407.

———, ———, ——— (1980) 'Britain', in D. Seers and C. Vaitsos (eds) *Integration and Unequal Development. The experience of the EEC*, Basingstoke: Macmillan, 136–54.

Finger, J.M. and A.J. Yeats (1976) 'Effective protection by transportation costs and tariffs: a comparison of magnitudes', *Quarterly Journal of Economics* 40: 169–76.

Flanagan, R.J. (1993) 'European wage equalization since the Treaty of Rome', in

L. Ulman, B. Eichegreen and W.T. Dicken (eds) *Labor and an Integrated Europe*, Washington: The Brookings Institution, 167–87.

Fothergill, S. and G. Gudgin (1982) *Unequal Growth. Urban and regional employment change in the U.K.*, London: Heinemann.

Fotheringham, A.S. and M.E. O'Kelly (1989) *Spatial Interaction Models: formulations and applications*, Dordrecht: Kluwer.

Francis, A. and P.K.M. Tharakan (eds) *The Competitiveness of European Industry*, London: Routledge.

General Agreement on Tariffs and Trade (GATT) (1992) *International Trade 90–91*, Vol. II, Geneva: GATT.

George S. (1990) *An Awkward Partner: Britain in the European Community*, Oxford: Clarendon Press.

Geraci, V.J. and W. Prewo (1977) 'Bilateral trade flows and transport costs', *Review of Economics and Statistics* 59: 67–74.

Gibb, R.A. (ed.) (1994) *The Channel Tunnel. A geographical perspective*, Chichester: Wiley.

Gibb, R.A., R.D. Knowles and J.H. Farrington (1992) 'The Channel Tunnel rail link and regional development: an evaluation of British Rail's procedures and policies', *Geographical Journal* 158: 273–85.

Giersch, H. (1949–50) 'Economic union between nations and the location of industries', *Review of Economic Studies* 17: 87–97.

Gordon, I.R. (1976) 'Gravity demand functions, accessibility and regional trade', *Regional Studies* 10: 25–37.

——— (1978) 'Distance deterrence and commodity values', *Environment and Planning* A, 10: 889–900.

——— (1985) 'Economic explanations of spatial variation in distance deterrence', *Environment and Planning* A, 17: 59–72.

Gordon I.R. and A.P. Thirlwall (eds) (1989) *European Factor Mobility. Trends and consequences*, Basingstoke: Macmillan.

Gould, A. and D.E. Keeble (1984) 'New firms and rural industrialization in East Anglia', *Regional Studies* 18: 189–201.

Grassman, S. (1980) 'Long-term trends in openness of national economies', *Oxford Economic Papers* 32: 123–33.

Greenaway, D., H. Hyclak and R. Thornton (eds) (1989) *Economic Aspects of Regional Trading Arrangements*, Brighton: Harvester Wheatsheaf.

Hanink, D.M. (1988) 'An extended Linder model of international trade', *Economic Geography* 64: 322–34.

Harris, R.I.D. (1989) *The Growth and Structure of the UK Regional Economy 1963–85*, Aldershot: Avebury.

Harrison, R.T. and M. Hart (eds) (1993) *Spatial Policy in a Divided Nation*, London: Jessica Kingsley.

Haynes, K.E. and A.S. Fotheringham (1984) *Gravity and Spatial Interaction Models*, Beverley Hills: Sage.

Hayuth, Y. (1992) 'Multimodal freight transport', in B.S. Hoyle and R.D. Knowles (eds) *Modern Transport Geography*, London: Belhaven Press, 199–214.

Healey, M.J. and B.W. Ilbery (eds) (1985) *The Industrialization of the Countryside*, Norwich: Geo Books.

Hirschman, A.O. (1958) *The Strategy of Economic Development*, New Haven: Yale University Press.

Hitchens, D.M.W.N., K. Wagner and J.E. Birnie (1990) *Closing the Productivity Gap: a comparison of Northern Ireland, the Republic of Ireland, Britain and West Germany*, Aldershot: Gower–Avebury.

——, ——, —— (1991) 'Northern Ireland's productivity failure: a matched plant comparison with West Germany', *Regional Studies* 25: 111–21.

——, ——, —— (1992) 'Competitiveness and regional development: the case of Northern Ireland', *Regional Studies* 26: 106–14.

Hoare, A.G. (1986) 'British ports and their hinterlands: a rapidly changing geography', *Geografiska Annaler* B, 68: 29–40.

—— (1988) 'Geographical aspects of British overseas trade: a framework and a review', *Environment and Planning* A, 20: 1345–64.

—— (1993) 'Domestic regions, overseas nations, and their interactions through trade: the case of the United Kingdom', *Environment and Planning* A, 25: 701–22.

Holliday, I., G. Marcou and R. Vickerman (1991) *The Channel Tunnel. Public policy, regional development and European integration*, Belhaven Press, London.

Höltgen, D. (1992) 'Güterverkehrszentren. Knotenpunkte des Kombinierten Verkehrs im europäischen Binnenmarkt', *Geographische Rundschau* 12/92: 708–15.

Ihde, G.B. (1991) *Transport, Verkehr, Logistik*, Munich: Verlag Franz Vahlen.

Isard, W. and D.F. Bramhall (1960) 'Gravity, potential, and spatial interaction models', in W. Isard *et al.*, *Methods of Regional Analysis: an introduction to regional science*, Cambridge, Mass.: MIT Press, 493–568.

Isard, W. and M.J. Peck (1954) 'Location theory and international and interregional trade theory', *Quarterly J. of Economics* 68: 97–114.

Jacquemin, A. and A. Sapir (eds) (1989) *The European Internal Market. Trade and competition*, Oxford: Oxford University Press.

Johnston, R.D. (1976) *The World Trading System. Some enquiries into its spatial structure*, London: Bell.

Jovanovic, M.N. (1992) *International Economic Integration*, London: Routledge.

Kaldor, N. (1972) 'The irrelevance of equilibrium economics', *Economic Journal* 82: 1237–55.

Keeble, D.E. (1984) 'The urban–rural manufacturing shift', *Geography* 69: 163–6.

—— (1989) 'Core-periphery disparities, recession and new regional dynamisms in the European Community', *Geography* 74: 1–11.

—— (1991a) 'De-industrialization, new industrialization pressures and regional restructuring in the European Community', in T. Wild and P. Jones (eds) *De-Industrialization and New Industrialization in Britain and Germany*, London: Anglo-German Foundation for the Study of Industrial Society, 40–65.

—— (1991b) 'Core-periphery disparities and regional restructuring in the European Community of the 1990s', in H. Blotevogel (ed.) *Europäische Regionen im Wandel*, Dortmund: Dortmunder Vertrieb für Bau- und Plannungsliteratur, 49–68.

Keeble, D.E., J. Offord and S. Walker (1986) 'Peripheral regions in a community of twelve member states. Final report', Department of Geography, University of Cambridge, for the Directorate-General for Regional Policy, Commission of the European Communities. (Published as Keeble *et al.* 1988.)

——, ——, —— (1988) *Peripheral Regions in a Community of Twelve Member States*, Luxembourg: Office for Official Publications of the European Communities.

——, P.L. Owens and C. Thompson (1981) 'The influence of peripheral and central locations on the relative development of regions', Final report to the Regional Policy Directorate-General, Commission of the European Communities, and the United Kingdom Department of Industry. (Published as *Centrality, Peripherality and EEC Regional Development*, London: HMSO, 1982.)

——, ——, —— (1982) 'Regional accessibility and economic potential in the European Community', *Regional Studies* 16: 419–31.

Keeble, D.E., P. Tyler, G. Broom and J. Lewis (1992) *Business Success in the Countryside. The performance of rural enterprise*, London: HMSO.

Kindleberger, C.P. (1956) *The Terms of Trade: a European case study*, New York: MIT Press and Wiley.

——— (1967) *Europe's Postwar Growth. The role of labor supply*, Cambridge, Mass.: Harvard University Press.

King, L.J. (1969) *Statistical Analysis in Geography*, Englewood Cliffs: Prentice-Hall.

Kravis, I.B. and R.E. Lipsey (1971) *Price Competitiveness in World Trade*, New York: National Bureau of Economic Research.

Krugman, P. (1980) 'Scale economies, product differentiation, and the pattern of trade', *American Economic Review* 70: 950–9.

Latham, A.J.H. (1978) *The International Economy and the Underdeveloped World 1865–1914*, London: Croom Helm.

Lin, S.A.Y. and M.A. Hanson (1976) 'Transportation sensitivity and regional growth', *Regional Science and Urban Economics* 6: 309–25.

Linneman, H. (1966) *An Econometric Study of International Trade Flows*, Amsterdam: North Holland.

——— (1969) 'Trade flows and geographical distance, or the importance of being neighbours', in H.C. Bos (ed.) *Towards Balanced International Trade*, Amsterdam: North Holland.

Lipsey, R.E. and M.Y. Weiss (1974) 'The structure of ocean transport charges', *Explorations in Economic Research* 1: 162–93.

Logan, R. (1971) 'Transport and communications in industrial mobility: the UK experience with particular reference to Scotland and Northern Ireland', Ph.D. thesis, University of Glasgow.

Lord Berkeley (1994) Personal communication.

Lord Cockfield (1988) 'Foreword', in P. Cecchini, *The European Challenge 1992. The benefits of a single market*. Aldershot: Wildwood House, xiii–xiv.

McConnell, J.E. (1986) 'Geography of international trade', *Progress in Human Geography* 10: 467–83.

Mackay, R.R. (1992) '1992 and relations with the EEC', in P. Townroe and R. Martin (eds) *Regional Development in the 1990s. The British Isles in transition*, London: Jessica Kingsley, 278–87.

McKinnon, A.C. (1992) 'Manufacturing in a peripheral location: an assessment of logistical penalties', *International Journal of Logistics Management* 3: 31–48.

Macmillan, M. (1982) 'The economic effects of international migration: a survey', *Journal of Common Market Studies* XX: 245–67.

Marshall, J.N., P. Wood, P.W. Daniels, A. McKinnon. J. Bachtler, P. Damesick, N. Thrift, A. Gillespie, A. Green and A. Leyshon (1988) *Services and Uneven Development*, Oxford: Oxford University Press.

Martin, R.L. (1988) 'The political economy of Britain's north–south divide', Institute of British Geographers, *Transactions* 13: 389–418.

——— (1993) 'Remapping British regional policy: the end of the North–South divide?', *Regional Studies* 27: 797–805.

Mayes, D.G. (1978) 'The effects of economic integration on trade', *Journal of Common Market Studies*, XVII (September): 1–25.

——— (ed.) (1993) *External Implications of European Integration*, Hemel Hempstead: Harvester Wheatsheaf.

de Melo, J. and A. Panagariya (eds) (1993) *New Dimensions in Regional Integration*, Cambridge: Cambridge University Press.

Mendes, A.J.M. (1987) *Economic Integration and Growth in Europe*, London: Croom Helm.

Millington, A.I. (1988) *The Penetration of EC Markets by UK Manufacturing Industry*, Aldershot: Avebury.

Ministry of Transport (1966) *Portbury. Reasons for the Minister's decision not to authorise the construction of a new dock at Portbury. Bristol*, London: HMSO.

Molle, W. (1990) *The Economics of European Integration (theory, practice, policy)*, Aldershot: Dartmouth.

Mowatt, A. (1994) Private communication.

Myrdal, G. (1957) *Economic Theory and Under-Developed Regions*, London: Duckworth.

Neven, D.J. (1990) 'EEC integration toward 1992: some distributional aspects', *Economic Policy* 10: 14–62.

Nevin, E. (1990) *The Economics of Europe*, Basingstoke: Macmillan.

O'Brien, R. (1992) *Global Financial Integration: the end of geography*, New York: Council on Foreign Relations Press.

Organization for Economic Co-operation and Development (OECD) (1968) *Ocean Freight Rates as Part of Total Transport Costs*, Paris: OECD.

Owen, N. (1983) *Economies of Scale, Competitiveness, and Trade Patterns within the European Community*, Oxford: Clarendon Press.

PEIDA (Planning and Economic Consultants) (1984) *Transport Costs in Peripheral Areas*, mimeo, Edinburgh and Henley-on-Thames: PEIDA.

Peschel, K. (1981) 'On the impact of geographic distance on the inter-regional patterns of production and trade', *Environment and Planning* A, 13: 605–22.

——— (1985) 'Spatial structures in international trade: an analysis of long term developments', *Papers of the Regional Science Association* 58: 97–111.

Potter, J. (1993) 'External manufacturing investment in a peripheral rural region: the case of Devon and Cornwall', *Regional Studies* 27: 193–206.

Prewo, W. (1974) 'A multinational interindustry gravity trade model for the European Common Market', Ph.D thesis, Johns Hopkins University, Baltimore.

Robson, P. (1980) *The Economics of International Integration*, London: Allen & Unwin.

Rowthorn, R.E. and J.R. Wells (1987) *De-Industrialization and Foreign Trade*, Cambridge: Cambridge University Press.

Scitovsky, T. (1958) *Economic Theory and Western European Integration*, London: Allen & Unwin.

Scottish Office (1981) 'Transport cost in Scottish manufacturing industries', *Scottish Economic Bulletin* 22: 27–37.

Secretary of State for Employment (1989) *Employment in the Docks: The Dock Labour Scheme*, Cm 664, London: HMSO.

Stopford, J.M. (1993) 'European multinational competitiveness: implications for trade policy', in D.G. Mayes (ed.) *The External Implications of European Integration*, Hemel Hempstead: Harvester Wheatsheaf, 52–78.

Tolley, R.S. and B.J. Turton (eds) (1987) *Short-Sea Crossings and the Channel Tunnel*, London: Transport Study Group, Institute of British Geographers.

Toothill, J.N. (1962) *Inquiry into the Scottish Economy*, Report of a Committee Appointed by the Scottish Council. Edinburgh: Scottish Council (Development and Industry).

Tovias, A. (1982) 'Testing factor price equalization in the EEC', *Journal of Common Market Studies* XX: 375–88.

Townroe, P. and R.L. Martin (eds) (1992) *Regional Development in the 1990s. The British Isles in transition*, London: Jessica Kingsley.

Tsoukalis, L. (1991) *The New European Economy. The politics and economics of integration*, Oxford: Oxford University Press.

Tyler, P. and M. Kitson (1987) 'Geographical variations in transport costs of manufacturing firms in Great Britain', *Urban Studies* 24: 61–73.

Tyler, P., B.C. Moore and J. Rhodes (1988a) 'Geographical variations in industrial costs', *Scottish Journal of Political Economy* 35: 22–50.

——, ——, —— (1988b) *Geographical Variations in Costs and Productivity in England*, London: HMSO.

United Nations (1990) *World Economic Survey 1990*, New York: United Nations.

Vickerman, R.W. (1989) 'After 1992 – the South East as a frontier region', in M. Breheny and P. Congdon (eds) *Growth and Change in a Core Region: the case of South East England*, London: Pion, 87–105.

Vickerman, R.W. and A.D.J. Flowerdew (1990) *The Channel Tunnel: the economic and regional impact*, London: Economist Intelligence Unit.

Viner, P. (1950) *The Customs Union Issue*, New York: Carnegie Endowment for International Peace.

Williamson, J. and A. Bottrill (1971) 'The impact of customs unions on trade in manufactures', *Oxford Economic Papers* 23: 323–51.

Wilson, A.G. (1967) 'A statistical theory of spatial distribution models', *Transportation Research* 1: 253–69.

—— (1971) 'A family of spatial interaction models, and associated developments', *Environment and Planning* 3: 1–32.

Winters, L.A. (1987) 'Britain in Europe: a survey of quantitative trade studies', *Journal of Common Market Studies* XXV: 315–35.

—— (ed.) (1992) *Trade Flows and Trade Policy after 1992*, Cambridge: Cambridge University Press.

Winters, L.A. and A. Venables (eds) (1991) *European Integration: trade and industry*, Cambridge: Cambridge University Press.

World Bank (1990) *World Development Report 1990*, New York: Oxford University Press.

—— (1993) *World Development Report 1992*, New York: Oxford University Press.

Yeates, M.H. (1968) *An Introduction to Quantitative Analysis in Economic Geography*, New York: McGraw-Hill.

Young, A.A. (1928) 'Increasing returns and economic progress', *Economic Journal* XXXVIII: 527–42.

Youngson, A.J. (1967) *Overhead Capital: a study in development economics*, Edinburgh: Edinburgh University Press.

AUTHOR INDEX

SUBJECT INDEX